Organizing for Information Warfare at the Geographic Combatant Commands

Lessons from the United States Central Command Joint Effects Process

CHRISTOPHER PAUL, MICHAEL SCHWILLE, STEPHEN WEBBER, ALYSSA DEMUS, ERIK VAN HEGEWALD

Prepared for the United States Central Command
Approved for public release; distribution is unlimited

For more information on this publication, visit **www.rand.org/t/RRA2270-1**.

Published by the RAND Corporation, Santa Monica, Calif.

Library of Congress Cataloging-in-Publication Data is available for this publication.

ISBN: 978-1-9774-1274-4

Cover: Olena_T/Getty Images.

About This Report

This report documents and evaluates a series of recent changes (2020–2023) to battle rhythm events and processes at U.S. Central Command (USCENTCOM). These revised processes enable the consideration of the informational aspects of operations and support better integration of informational and physical effects across the spectrum of competition and conflict within the command. The report provides an evaluative framework useful for ensuring that future changes in battle rhythm events preserve the benefits gained to date and for guiding changes at other commands that want to gain similar benefits.

The research reported here was completed in August 2023 and underwent security review with the sponsor before public release.

RAND National Security Research Division

This research was sponsored by USCENTCOM and conducted within the International Security and Defense Policy Program of the RAND National Security Research Division (NSRD), which operates the National Defense Research Institute (NDRI), a federally funded research and development center sponsored by the Office of the Secretary of Defense, the Joint Staff, the Unified Combatant Commands, the Navy, the Marine Corps, the defense agencies, and the defense intelligence enterprise.

For more information on the RAND International Security and Defense Policy Program, see www.rand.org/nsrd/isdp or contact the director (contact information is provided on the webpage).

Acknowledgments

Thanks to COL (retired) Andrew Whiskeyman, former USCENTCOM J-39, for conceiving this project and to COL Sean Heidgerken, USCENTCOM J-39, and his staff (including Louis Morgan, Maj. Trent Schill, and Ted Dimone) for their support, assistance, input, and oversight as we completed the research. We are also indebted to the current and former USCENTCOM personnel and personnel at the regional service component commands who provided input and data during the project. The terms of our interviews prevent us from thanking you by name, but you know who you are, and you are appreciated! The efforts of RAND communication analyst David Adamson significantly improved the prose, presentation, and organization of this report—thank you. We further thank Maria Falvo for her support managing notes, citations, and formatting for this report, and Daphne Rozenblatt who rendered this report into its final form. We appreciate the contribution of RAND quality assurance reviewers Sherrill Lingel and Stephen Dalzell for their thoughtful comments on an earlier draft.

Summary

Information power is increasingly recognized as a critical source of national power and a key component of military operations. Between 2020 and 2023, U.S. Central Command (USCENTCOM) made changes to better integrate information and information warfare issues into its battle rhythm. *Battle rhythm* refers to the day-to-day planning, coordination, and integration of a geographic combatant command (GCC). These activities rely on cross-functional staff integration groups, referred to as boards, bureaus, centers, cells, and working groups (B2C2WG) and operational planning teams (OPTs). The meetings and processes generated by the B2C2WG are part of a command's routine cycle of command and staff activities that synchronize current and future operations (CUOPS and FUOPS).

As a result of these changes, several USCENTCOM operations have more fully integrated information and information power into planning and execution. These operations offer opportunities for assessment and learning, either as positive or negative examples. However, the extent to which these operations succeeded in integrating information as a joint function has not previously been well documented. Although USCENTCOM now has important experience integrating information into its battle rhythm, the effects have not been fully captured.

To address this gap, this report documents the processes followed and the information activities undertaken as part of these operations and assesses their effects in the information environment (IE). In addition, the report includes lessons relevant to other elements within the joint force and presents a framework for drawing further lessons from the commander's decisionmaking cycle and for the execution of information as a joint function.

To accomplish these goals, the research team employed several methods: a comprehensive review of relevant documentation and literature, both internal and external to the command; more than 35 interviews with subject-matter experts drawn from current and previous USCENTCOM and CENTCOM service component command (SCC) personnel; and case studies of three instances of operations planned under the USCENTCOM joint effects process (JEP).

Key Findings

RAND's review of literature, semistructured interviews with USCENTCOM staff, case studies, and observations of battle rhythm events showed that the revised JEP has improved USCENTCOM's ability to operate in the IE. Operating in the IE is inherently difficult, and the JEP remains a work in progress. Significant challenges remain to USCENTCOM's aspirations of conducting seamless effects-based operations across cognitive, informational, and physical spaces, but improvement has been noted across all areas as the staff continues to implement and refine the JEP.

U.S. Central Command Has Made Substantial Progress in Integrating Information Through the Adoption of the Joint Effects Process

The processes adopted between 2020 and 2023 engage senior leaders; centralize and elevate planning where appropriate at the GCC level; align operations, activities, and investments (OAI) to strategic objectives; layer these OAI and effects to achieve objects; promote information sharing across the command; bridge planning between CUOPS and FUOPS; and integrate special activities, among other things.

The positive attributes of the JEP helped USCENTCOM move from a reactive posture in the IE to a proactive one: In 2020, information was largely an "add-on" to operations. In 2023, it is a central consideration. The processes adopted have brought coherence to the OAI that USCENTCOM conducts, and component commands are now better aligned to the GCC and with one another.

We identified and developed a list of 19 characteristics of a JEP that contribute to the integration and effectiveness of information efforts. These characteristics support four major themes: An effective JEP

- (C) centralizes and elevates information power and related effects
- (A) aligns activity and strategy
- (I) integrates, synchronizes, synergizes, and layers informational and physical power
- (F) provides feedback and promotes iterative improvement.

USCENTCOM has made progress toward incorporating each of these characteristics across these four themes into the command battle rhythm. It

- (C) centralizes and elevates information power and related effects
  - (C1) receives significant senior leader attention
  - (C2) centralizes and elevates planning rather than leaving planning to the service components
  - (C3) elevates prominence of information in command processes and in SCC awareness; broadens staff awareness across the range of kinetic and nonkinetic OAI
  - (C4) involves the right people, in the right numbers
- (A) aligns activity and strategy
  - (A1) promotes a campaigning mindset
  - (A2) aligns and links OAI with command objectives
  - (A3) generates unified, synchronized, justified, and well-timed force planning requirements
  - (A4) demands justification of habitual and routinized events; takes compulsory events and layers additional OAI to support them in ways that better tie to theater campaign and intermediate military objectives (IMOs)
  - (A5) supports or bridges a medium-scale planning time horizon (between current and future operations)

- (I) integrates, synchronizes, synergizes, and layers informational and physical power
  - (I1) is able to generate concepts of operations that support multiple objectives or levels; able to coordinate multiple OAI against a single effect or objective
  - (I2) deconflicts and coordinates authorities and permissions across the service components and other elements
  - (I3) promotes or enables information sharing across relevant staff sections, components, and capabilities
  - (I4) integrated with intelligence such that process is intelligence driven and encourages additional intelligence support
  - (I5) accounts for and integrates special activities
- (F) provides feedback and promotes iterative improvement
  - (F1) is able to provide effective feedback and input to the theater campaign plan and theater campaign order (TCO) process and promote clear IMOs
  - (F2) supports dynamic and evolving objectives between iterations of TCO cycle; responsive to emergent events
  - (F3) supports assessment of operational effectiveness
  - (F4) supports assessment of campaign effectiveness; tracks actual trajectory of OAI against TCO projections
  - (F5) is able to track past and planned OAI, balance between retrospective and prospective view.

These 19 characteristics can serve as benchmarks and targets for progress toward information integration or as targets for other commands that wish to improve integration of information with other command activities.

The Joint Effects Process Is a Work in Progress, and Areas for Improvement Remain to Fully Integrate Information at U.S. Central Command

While the JEP has helped USCENTCOM to move toward an effects-based approach that better incorporates information, there is still work to be done. Operating in the IE is inherently difficult, so the JEP will be a work in progress for the foreseeable future. USCENTCOM faces three general challenge areas that prevent the command from reaching its aspirations for effects-based operations in the IE. These are assessment, intelligence, and personnel.

Assessing information effects is generally more difficult than assessing strictly physical effects, and insufficiently specific higher-level objectives compounds this difficulty. Additionally, there are challenges related to integrating intelligence into the JEP. As USCENTCOM reorients itself after two decades of in-theater conflict, its use of intelligence must also shift its primary focus from kinetic strikes to understanding target audiences' behaviors and helping planners define effects and how to measure them. Furthermore, the USCENTCOM JEP requires nontraditional skill sets and modes of thinking, and USCENTCOM will need to work hard to make sure that assigned personnel acquire and maintain the necessary expertise.

Recommendations

This report offers several recommendations to USCENTCOM going forward:

- Sustain the JEP while continuing to refine and improve processes.
- Continue to review and challenge habitual activities while continuing to work to connect OAI to the TCO and IMOs.
- Develop and implement a method for articulating risk of action versus risk of inaction.
- Increase process discipline by adopting an effects template.
- Improve assessment by continuing to improve the assessability of IMOs.
- Improve assessment by better defining supporting intelligence collection requirements.
- Improve assessment by considering objectives and related assessment requirements across short-, medium-, and long-term time horizons.
- Sustain and expand the use of command and control in the information environment (C2IE) software for event tracking and reporting.
- Integrate key leader engagement as one of the capabilities that is tracked, coordinated, and deconflicted through the JEP.
- Employ the 19 criteria from the framework for the evaluation and comparison of battle rhythm events developed in this report to continue to improve the JEP and to help ensure that future evolutions of B2C2WG processes do not move away from desirable characteristics achieved during the study time frame.

Contents

Figures and Tables

Figures

Tables

CHAPTER 1

Introduction

Awareness of the importance of informational power as a source of national power and its relevance in military operations across the spectrum of conflict has been ascendant within the U.S. Department of Defense (DoD) for some time.[1] To improve the inclusion, integration, planning, and execution of information as a joint function and operations in the information environment (OIE) in command operations, U.S. Central Command (USCENTCOM) made a series of changes to elements and processes within the command's battle rhythm events between 2020 and 2023. These changes were focused within working groups (WGs) in the command's joint effects process (JEP), specifically the influence effects working group (IEWG), the joint effects working group (JEWG), the joint effects board (JEB), the special activities working group (SAWG), the joint targeting working group (JTWG), and the planning and resource alignment conference (PRAC).

Battle rhythm refers to the day-to-day work of planning, coordination, and integration within a geographic combatant command (GCC), which relies on cross-functional staff integration elements or groups, referred to in the aggregate as boards, bureaus, centers, cells, and working groups (B2C2WG) and operational planning teams (OPTs).[2] The meetings and processes generated by the B2C2WG are part of a command's routine cycle of command and staff activities intended to synchronize current and future operations (CUOPS and FUOPS).[3]

In this report, we document the processes that are followed and the information activities that are undertaken as part of these operations. In addition, we distill lessons related to the characteristics leading to effectiveness in the USCENTCOM JEP, providing a framework to help USCENTCOM to sustain its progress and continue to improve. This framework should be useful for other commands that seek the same sorts of benefits that have been accrued by USCENTCOM.

[1] Christopher Paul, Michael Schwille, Michael Vasseur, Elizabeth M. Bartels, and Ryan Bauer, *The Role of Information in U.S. Concepts for Strategic Competition*, RAND Corporation, RR-A1256-1, 2022.

[2] Joint Staff J7, *Insights and Best Practices Focus Paper: Joint Headquarters Organization, Staff Integration, and Battle Rhythm*, 3rd ed., September 2019.

[3] *Battle rhythm* is described as follows: "The battle rhythm provides structure and sequencing of actions and events within the HQs [headquarters] regulated by the flow and sharing of information supporting all decision cycles" (Joint Publication 3-33, *Joint Task Force Headquarters*, U.S. Joint Chiefs of Staff, January 31, 2018).

The research addressed the following questions:

- What are the details of the revised processes and staff organizational architecture?
- What do the revised battle rhythm events deliver for the commander, especially in terms of the integration and effectiveness of information-related operations, activities, and investments (OAI)?
- How effectively does the change in battle rhythm perform in operations, and how does that compare with performance prior to its adoption?
- To the extent that performance improves, why is this the case? What characteristics or features of the new JEP contribute to its effectiveness?

Answers to these questions support our research's overall objective to (1) offer lessons for the improvement of future doctrine, planning, management, resourcing, execution, and assessment for integrating OIE with broader operations at USCENTCOM and across the joint force, and (2) provide a template for future analogous analyses given that most GCC, joint force, and service entities have struggled with institutionalizing and operationalizing information warfare.[4]

Background and Problem

The recent emphasis on information and OIE has been prominent in joint and service doctrine and concepts and noted repeatedly in senior leader statements.[5] As is always the case with major organizational or cultural changes, the adoption and embrace of information and OIE proceed at different paces within different elements and at different echelons within DoD.[6]

From 2020 through 2023, USCENTCOM adopted changes to reflect this emphasis. The commander and senior staff increased the command's emphasis on and integration of information because of the increasing prominence of OIE in doctrine, concepts, and strategic guidance, as well as other changes in the command and within its area of responsibility

---

[4] See, for example, Paul et al., 2022.

[5] Consider, for example, the adoption of information as one of the joint functions in Joint Publication 1-0, *Joint Personnel Support*, U.S. Joint Chiefs of Staff, December 1, 2020; U.S. Joint Chiefs of Staff, *Joint Publication Joint Concept for Operating in the Information Environment (JCOIE)*, July 25, 2018; Joint Publication 3-04, *Information in Joint Operations*, U.S. Joint Chiefs of Staff, September 14, 2022; Marine Corps Doctrinal Publication 8, *Information*, U.S. Department of the Navy, June 21, 2022.

[6] To see the addition of information as a joint function described as a paradigm shift, see Scott K. Thomson and Christopher E. Paul, "Paradigm Change: Operational Art and the Information Joint Function," *Joint Force Quarterly*, No. 89, 2nd Quarter 2018. For a discussion of related cultural and organizational change, see Christopher Paul and Isaac Porche III, "Toward a U.S. Army Cyber Security Culture," *International Journal of Cyber Warfare and Terrorism*, Vol. 1, No. 3, July–September 2012.

(AOR). These changes included the end of major combat operations in the AOR, changes in the national military strategy (including an increased willingness to accept risk in the USCENTCOM AOR in favor of other strategic priorities), changes in available resources, and a DoD-wide desire to increase emphasis on partners.[7] Collectively, these pressures led to the changes in battle rhythm events documented and analyzed in this report.

In that time span (2020–2023), changes in battle rhythm processes and structures led to several USCENTCOM operations that have been noteworthy for the way in which information and information power have been integrated in both planning and execution.[8] These operations have all offered opportunities for insight, analysis, assessment, and learning—either as positive or negative examples.

Unfortunately, the extent to which inclusion, integration, planning, and execution of information as a joint function succeeded in these operations has not been well documented. Although USCENTCOM has important experience integrating information as a joint function into operations, observations, lessons, and assessments of these efforts, its battle rhythm processes, execution, and effects were only partially captured by the command history process, which risked the loss of relevant lessons. This research is intended to document these changes and assess their effects.

Methods and Approach

To meet the analytic needs of USCENTCOM and document the changes and characteristics of their reformed battle rhythm events, this research employed several methods. Each is described briefly below.

Literature Review

The foundation of this research is a literature and document review. Building on our previous research and experience related to organizing for and conducting OIE and the information joint function, we identified relevant existing studies, histories, doctrine, white papers, and reports.[9] After reviewing relevant published concepts and scholarship, we turned to the

---

[7] Senior military officers with expertise in OIE and experience with USCENTCOM battle rhythms and priorities, interview with the authors via videoconference, March 3, 2023.

[8] The 2018 Joint Concept for OIE uses the term *informational* to "reflect the way that individuals, information systems, and groups communicate and exchange information. Informational aspects are the features and details of activities that an observer interprets and uses to assign meaning." We use the term *informational power* to describe the ability to understand and shape these aspects of an activity or event. See U.S. Joint Chiefs of Staff, 2018.

[9] Examples of the authors' relevant previous work include Paul et al., 2022; Michael Schwille, Jonathan Welch, Scott Fisher, Thomas M. Whittaker, and Christopher Paul, *Handbook for Tactical Operations in the Information Environment,* RAND Corporation, TL-A732-1, 2021; Michael Schwille, Anthony Atler, Jonathan Welch, Christopher Paul, and Richard C. Baffa, *Intelligence Support for Operations in the Information*

archival documents available within USCENTCOM: briefings, notes, slides, after-action reports, and other artifacts of various battle rhythm events and supported OAI.

Interviews

We also conducted more than 35 interviews with current and former USCENTCOM uniformed, civilian, and contractor personnel, as well as with personnel from regional component commands with current or former roles in the studied battle rhythm events. These interviews served several purposes, including:

- providing context to the various USCENTCOM documents and battle rhythm artifacts reviewed
- furnishing accounts of the history and evolution of the adoption of the reformed JEP
- bringing forth firsthand accounts and experiences with the studied battle rhythm events
- revealing the specific changes wrought by the new processes and the impact of those changes
- illuminating the operations supported by the new battle rhythm events.

Case Studies

To explore the effectiveness of the changes to USCENTCOM's battle rhythm events, we conducted a case-study analysis of three sets of USCENTCOM operations, one undertaken wholly before the adoption of the revised processes and two with events and elements that occurred before, during, and after the adoption of the revised JEP. These case studies were built with data drawn from USCENTCOM documentation (part of the literature review), documentation produced by other commands and joint entities, and input from personnel involved in the planning and execution of related OAI (as part of our interviews). Because the operational details of the case studies are sensitive in some instances, we provide case documentation and analysis in a supporting limited-distribution document. Top-level takeaways from that analysis are sufficient to support the other analyses here. We found that the level of integration of information with other efforts and the overall effectiveness of information as a part of operations were much greater after the adoption of the revised JEP process.

Environment: Dividing Roles and Responsibilities Between Intelligence and Information Professionals, RAND Corporation, RR-3161-EUCOM, 2020; Christopher Paul, Colin P. Clarke, Bonnie L. Triezenberg, David Manheim, and Bradley Wilson, *Improving C2 and Situational Awareness for Operations in and Through the Information Environment*, RAND Corporation, RR-2489-OSD, 2018; Alyssa Demus, Elizabeth Bodine-Baron, Caitlin McCulloch, Ryan Bauer, Christopher Paul, Jonathan Fujiwara, Benjamin J. Sacks, Michael Schwille, Marcella Morris, and Kelly Beavan, *Operationalizing Air Force Information Warfare*, RAND Corporation, RR-A1740-1, forthcoming.

Framework for Evaluating and Comparing Battle Rhythm Events

Building on the successful aspects of the revised USCENTCOM JEP, we created a framework that captures the desirable characteristics of a JEP, compares prior USCENTCOM battle rhythm events with the revised process, and evaluates effects processes at other GCCs or helps sustain desirable characteristics as USCENTCOM's processes undergo further evolution under different commanders.[10] The framework (presented in greater detail in Chapter 3) uses insights gained through this research and other efforts. The framework builds on our previous experience with OIE-related staffing processes and arrangements, insights gleaned from the interviews, and analyses of the case studies. Working independently, several authors compiled candidate lists of process characteristics favorable to the observed and reported beneficial outcomes. These were then reconciled into a single list, which was validated through a presentation to stakeholders who were previously interviewed as part of the research. For this presentation, we sorted these characteristics and features into four thematic categories.

We then applied the framework (Chapter 4) to assess the revised USCENTCOM JEP and compare it with the command's prior set of battle rhythm events. We undertook a thematic analysis by binning data from interviews, observations, and case studies according to the corresponding attributes in the framework. This application of the framework shows which USCENTCOM process changes have led to improvements and where there is room for continued improvement, as well as offers guidelines for future evolutions of the command's battle rhythm events so that beneficial characteristics and features can be retained.

Layout of the Remainder of the Report

In Chapter 2, we describe the various battle rhythm events that compose the USCENTCOM JEP. In Chapter 3, we present the details and the development process for the framework we used to identify desirable characteristics of a JEP. In Chapter 4, we apply this framework and assesses the revised USCENTCOM JEP against earlier processes. Chapter 5 compiles and summarizes findings and recommendations.

---

[10] By *desirable characteristics*, we denote characteristics that contribute to the JEP's ability to integrate informational and physical power and include consideration of OIE in support of the command's objectives.

CHAPTER 2

U.S. Central Command Joint Effects Process Planning and Battle Rhythm Events

USCENTCOM made many adjustments to its battle rhythm during the 2020–2023 period, during which many activities were discontinued, while new processes and battle rhythm events were established. In this chapter, we describe the USCENTCOM JEP that took shape during this time frame and the key battle rhythm events of that process, especially regarding the IEWG, the JEWG, the JEB, the SAWG, the JTWG, and the PRAC. The discussion also notes the purpose of each battle rhythm event and draws insights related to its functioning and performance from our literature review and interviews.

General Discussion on Planning and the Joint Effects Process

While most units and commands use processes outlined in doctrine and staff structures that have long been institutionalized, the types of operations that unit and HQ staff plan can vary widely. One of the main points of differentiation is the breadth of their *decision cycle*, which refers to the staff's ability to sense their environment, process information, and use it to enable decisionmaking by the commander. The commander then makes decisions and provides direction and guidance back into the system.[1] The commander's intent is then translated into plans and orders, which are disseminated to subordinate units for follow-on planning and execution. As one moves up the echelons from tactical units to joint HQ and GCCs (such as USCENTCOM), the decision cycle follows a similar model but expands outward in scale and scope. The time frames that the staff are responsible for considering become longer, and the types of issues they must address become more complex. A service component staff, for example, conducts more operational planning than GCC HQ, which addresses resourcing via the global force management and allocation process (GFMAP) and the policy, program-

[1] For doctrinal references on planning and operations, see Joint Publication 5-0, *Joint Planning*, U.S. Joint Chiefs of Staff, December 1, 2020; Joint Publication 3-0, *Joint Operations*, U.S. Joint Chiefs of Staff, incorporating change 1, October 22, 2018. For a theoretical view of command and control, see John Boyd, "An Organic Design for Command and Control," in Grant T. Hammond, ed., *A Discourse on Winning and Losing*, Maxwell Air Force Base, March 1, 2018.

ing, budgeting, and execution (PPB&E) process while overseeing the operations conducted across its entire AOR by service component commands (SCCs).

The process by which commands process information—the series of B2C2WG that work and meet on an established schedule in relation to other events—is known as its *battle rhythm*. USCENTCOM's battle rhythm is the focus of this section. As discussed in the introduction, the command has implemented and continues to refine the JEP. The JEP is intended to facilitate information flow within the command so that the USCENTCOM commander can issue guidance that turns into coherent plans and orders for the GCC's subordinate HQ. The various things that USCENTCOM—and its subordinate units—do across the theater are known as *OAI*. This term is a catch-all for what military forces do. Ideally, the battle rhythm helps USCENTCOM conduct OAI across the AORs that are aligned to the commander's intent.

USCENTCOM's battle rhythm functions across multiple, overlapping decision cycles. For example, the USCENTCOM J5 directorate, which proposes strategies, plans, and policy recommendations to the Chairman of the Joint Chiefs of Staff, thinks approximately one year into the future.[2] The USCENTCOM J3 plans section's horizon is a maximum of one year out and is considered in the J35 Future Plans section. Given the high operational tempo USCENTCOM often deals with, the FUOPS cell is often called on to plan as closely as 72-hours in the future, though it prefers to look further into the future.[3] The USCENTCOM J33 CUOPS team is generally concerned with things occurring within the same week.[4] The JEP is intended to align these overlapping decision cycles to feed a broader process that allows commanders to understand the environment and make decisions. Ideally, the decisions USCENTCOM makes are grounded in the principle of effects-based operations.[5] This means, as the term suggests, that OAI are undertaken to achieve a specific outcome, and everything happening across the USCENTCOM AOR is grounded in desired effects that support higher-level objectives. How this occurs will be further detailed in the next section.

Joint Effects Process

The JEP is what USCENTCOM uses to align SCC OAI, resources, plans, timelines, and desired effects. The process aligns multiple WGs, planning efforts, and USCENTCOM activities throughout a yearlong cycle. While the JEP is the overarching process, it comprises three separate yet connected subprocesses. These main processes can be categorized as (1) influence and effects, (2) special activities, and (3) targeting. These three subprocesses in turn

[2] USCENTCOM civilian with expertise in OIE and planning, in-person interview with authors, September 15, 2022.

[3] USCENTCOM civilian with expertise in OIE and planning, in-person interview with authors, September 15, 2022.

[4] Military officer with expertise in planning, in-person interview with the authors, December 5, 2022.

[5] For a classic discussion of effects-based operations, see Paul K. Davis, *Effects-Based Operations (EBO): A Grand Challenge for the Analytical Community,* RAND Corporation, MR-1477-USJFCOM/AF, 2001.

comprise a number of additional boards and WGs. These additional B2C2WG primarily consist of WGs that act to provide specific analysis, coordinate efforts across staff sections, and provide information for the three subprocesses we detail in the remainder of this chapter. As USCENTCOM evolved the JEP, many of its B2C2WG have gone defunct or were replaced by new activities. The current lowest-tier WGs are

- Cyber Effects WG
- Inter-Agency WG
- Key Leader WG
- Military Information Support Operations (MISO) Effects WG (MEWG)
- Operations Security (OPSEC) WG
- Joint Effects Electromagnetic WG
- Special Technical Operations (STO)
- Target Development WG.

We do not detail these additional subordinate processes and instead focus solely on the three major subprocesses. Each of the three major subprocesses will be further detailed throughout the remainder of this chapter.

The JEP starts with strategic guidance as articulated by the USCENTCOM commander and focuses on the effects that the SCCs and staff will prioritize. This guidance is published twice a year at the USCENTCOM component commanders' conference, a venue in which all the SCC commanders gather to advocate for service priorities and discuss both concluded and upcoming OAI. Guidance into the process also comes from refinements made at the JEB.

The JEP operates on a six-week cycle, with five of the weeks focused on regular WG activities and the sixth week on convening the JEB.[6] Figure 2.1 shows the joint effects cycle and weekly focus.

Upon receipt of the guidance, several processes are initiated to plan and align activities that include the IEWG and the JEWG. The IEWG looks the furthest into the future, primarily planning and aligning activities roughly nine months into the future. Its primary purposes are to identify opportunities to schedule new OAI, propose new strategic guidance, provide concept sketches for future activities, and provide inputs to the published effects order.[7] Outputs from the IEWG are used as inputs for the JEWG, which primarily plans between 90 to 180 days out from the execution of an OAI. The JEWG helps to build a common operating picture, provides initial OAI assessments, and focuses on planning 180 days prior to the exe-

[6] The six WGs are the same group that meets weekly, not six separate groups.

[7] Military officers with expertise in planning and operations, in-person interview with the authors, September 14, 2022.

FIGURE 2.1
Joint Effects Cycle

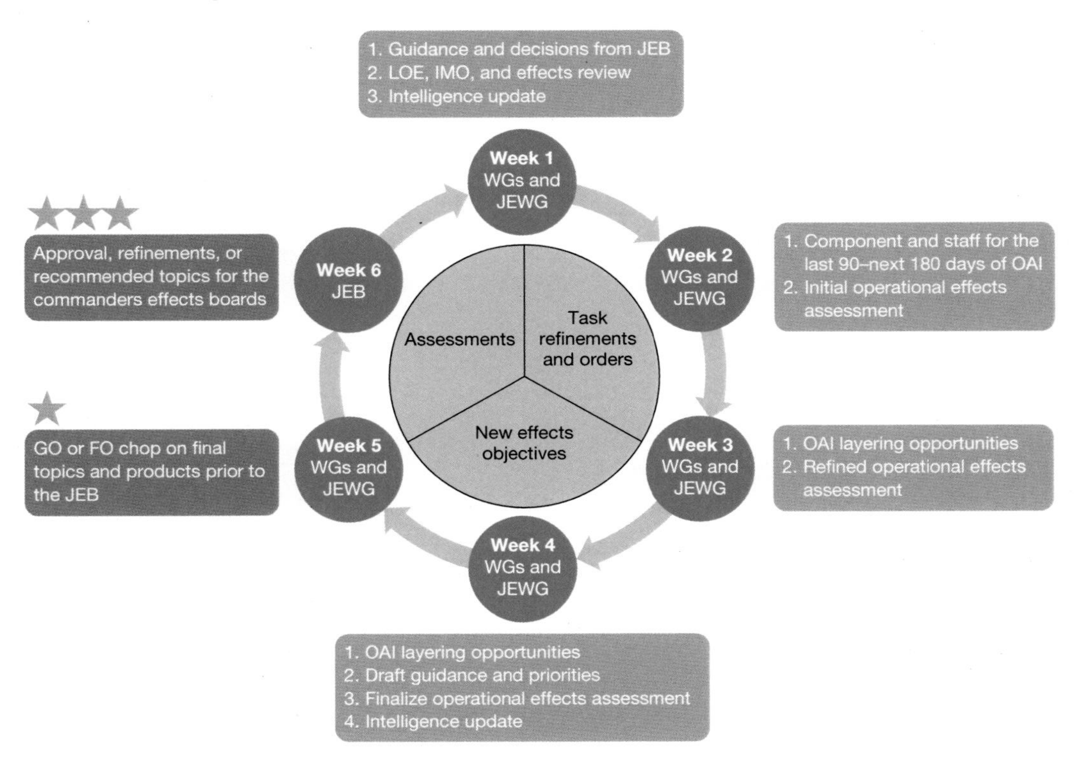

SOURCE: Adapted from USCENTCOM slide, USCENTCOM-provided briefing, October 27, 2021.
NOTE: FO = Flag Officer; GO = General Officer; IMO = intermediate military objective; JEB = joint effects board; JEWG = joint effects working group; LOE = lines of effort; OAI = operations, activities, and investments; WG = working group. Stars indicate the rank or grade of the general or flag officer who regularly chairs the meeting at that point in the cycle.

cution of an OAI through 90 days after execution. An operational effects assessment is the main product from this WG.[8]

The JEB focuses on the synchronization of lethal and nonlethal effects objectives. It approves the work conducted in the IEWG and JEWG, approves OAI, integrates effects across time horizons, and makes resourcing decisions.[9] Outputs from the JEB are used in the J5 directorate's theater campaign assessment and act to refine the commander's guidance. Additionally, the JEB informs J5 plans, USCENTCOM LOEs, and focused WGs and provides guidance for the PRAC, which is the activity that aligns resources to SCCs across a three-year

[8] Civilian with expertise in assessments, in-person interview with authors, December 6, 2022; Military officers with expertise in planning, in-person interview with the authors, December 6, 2022.

[9] Military officers with expertise in planning, in-person interview with the authors, December 6, 2022.

FIGURE 2.2
The U.S. Central Command Joint Effects Process

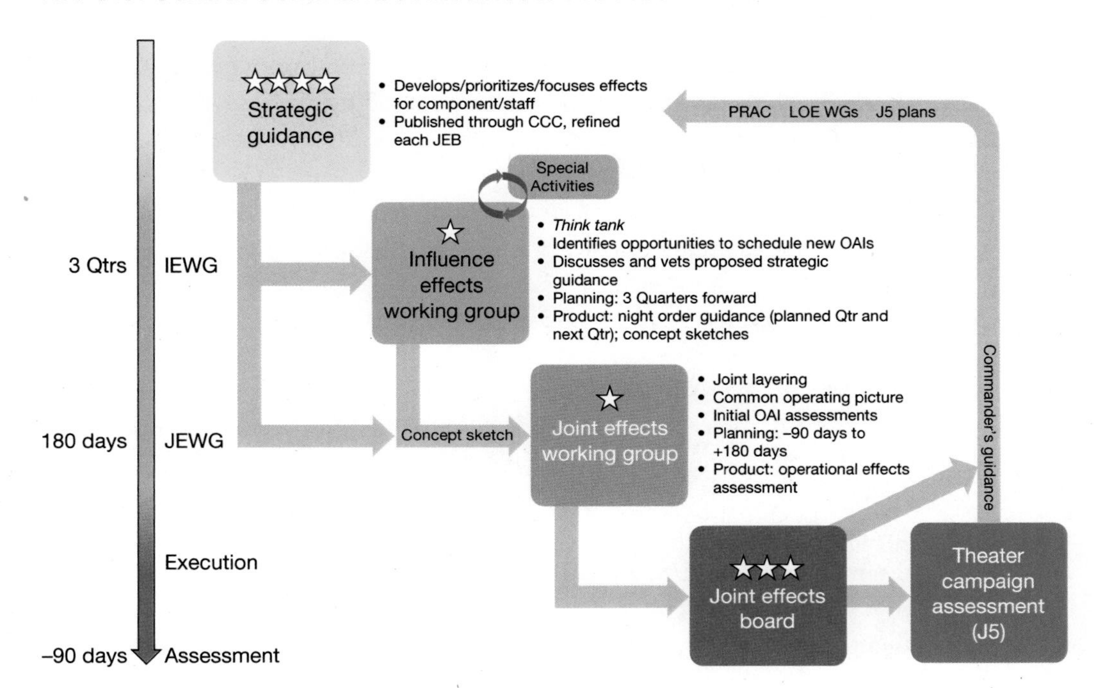

SOURCE: Adapted from a USCENTCOM-provided briefing, undated.
NOTE: CCC = Combatant Commander Conference; IEWG = influence effects working group; JEB = joint effects board; JEWG = joint effects working group; LOE = line of effort; OAI = operations, activities, and investments; PRAC = planning and resources alignment conference; Qtr = Quarter; WG = working group.

time horizon.[10] Figure 2.2 graphically represents the USCENTCOM JEP, highlights planning time horizons on the left-hand side of the figure, and shows how the process is cyclical.

How U.S. Central Command Achieves Effects in and Through the Information Environment

Effects are grounded in the conditions that USCENTCOM wishes to bring about. These are ideally derived from the objectives USCENTCOM hopes to achieve. Many of the desired effects are informational effects that affect the information available to certain targets or audiences or affect their perceptions, cognitions, or feelings, which ultimately affect their choices and behaviors. If an adversary is deterred from aggression, a partner is assured of the U.S. military's commitment to the region; if a terrorist group goes to ground out of fear that their network has been penetrated, it likely has to do with their perception and understand-

[10] Military officer with expertise in assessment, in-person interview with the authors, September 16, 2022.

ing of the situation, which is subject to informational effects. Every physical action—whether it be dropping a bomb on a terrorist leader, deploying military forces to the region, or meeting with a key leader—is seen and understood by various audiences. That information is received through the information environment (IE). The IE is the link between things that happen in physical domains and how they are perceived and understood by different audiences. The IE includes everything that shape this transition: how something is seen visually, written about in print, reported on television, discussed on social media, etc.[11] The JEP is intended to plan for the delivery of informational effects, including not only the transmission, communication, and perception of events and messages but also the understanding and meaning assigned by the receiver. In this way, each OAI is theoretically grounded not just in what it is (a military deployment, an exercise, or a kinetic strike) but what it is trying to achieve in and through the IE.

Traditional military processes are not always conducive to this kind of planning. Information is often treated as an *enabler*—something that is added to support other activities (which are generally physical actions, such as troops using loudspeakers to communicate to populations on the ground or dropping leaflets and sending radio transmissions to demoralize adversary troops). On staffs, information can be treated as an afterthought, something that is a late addition to a plan, such as an extra annex to a concept of operations (CONOPS) that addresses mostly physical activity.[12] As discussed in Chapter 1, USCENTCOM is experiencing a paradigm shift in how it plans and conducts operations. Rather than thinking primarily about physical effects (such as whether a missile destroys its target), it is now trying to elevate information effects (such as how that same missile strike is communicated and understood by different audiences) in its battle rhythm.[13] This entails modifying the processes by which USCENTCOM plans operations so that OAI are effects-based and both information and physical effects are fully considered. Within USCENTCOM, informational effects are conceptualized at the lowest-tier WG and act as inputs into the IEWG, SAWG, and JTWG.

While most of the physical activities are planned and conducted at lower echelons, USCENTCOM does retain several capabilities that are designed specifically to achieve effects in and through the IE. This creates a bias toward using these capabilities because they are available for immediate tasking with minimal bureaucratic hurdles. An example of such a capability is the Joint Military Information Support Operations Web Operations Center (JMWC). Unlike the SCCs, joint task forces (JTFs), or other task forces (TFs) in the

[11] For a conceptual overview of information warfare as a discipline, see David S. Alberts, John J. Garstka, Richard E. Hayes, and David A. Signori, *Understanding Information Age Warfare*, CCRP, 2001.

[12] For an overview of military planning in complex environments, see Joint Staff J7, *Insights and Best Practices Focus Paper: Design and Planning*, 1st ed., July 2013. For a discussion on how information is thought about, discussed, and incorporated into military planning, see Christopher Paul, "Is It Time to Abandon the Term Information Operations?" *Strategy Bridge*, March 11, 2019.

[13] Senior military officers with expertise in OIE and planning, in-person interview with the authors, December 5, 2022.

USCENTCOM AOR that might need to move personnel or capabilities to achieve an effect, the JMWC has several programs that can engage immediately with select audiences in the AOR. The evolution of the internet has created an opportunity to interact with and influence target audiences globally. This means that proximity to audiences is not necessarily a requirement.[14] While proximity might not be a problem, having a detailed understanding of target audiences, being available to closely monitor content, and being able to quickly react to that content are issues for USCENTCOM. As USCENTCOM priorities shift, having the capability to rapidly shift focus and message new audiences is critical. As one interviewee noted, this is the reason that the web operations (WebOps) capability should be retained at USCENTCOM.[15] However, this capability has inherent risks. Problems with the manner in which WebOps operates have caused the Office of the Under Secretary of Defense for Policy to request tighter controls on this capability. Whereas in the past, WebOps used to be approved at the level of the USCENTCOM commander, approvals appear to now be held at higher levels. This structural change will require more specific CONOPS to be approved further in advance for WebOps programs to have the ability to respond during crisis.[16]

Battle Rhythm Events

The JEP is made up of multiple B2C2WG. This study focused on five of them that encompass the most significant changes made to USCENTCOM's battle rhythm between 2020 and 2023. These are the IEWG, the JEWG, the SAWG, the JTWG, and the PRAC. Table 2.1 lists the main battle rhythm events and their key attributes.

Each one of these events operates for a specific purpose and on a different time horizon. Each one feeds information into the others, which results in CONOPS presented for leadership decision. Those CONOPS ideally shape OAI or add to OAI that are already underway. We reviewed documentation related to these events, observed selected events (the JEWG and the JEB), and conducted structured interviews with USCENTCOM subject-matter experts (SMEs) to understand what each event is, how it came into being, and how it currently functions.

[14] Civilians with expertise in social media activities and planning, in-person interview with the authors, December 6, 2022.

[15] Civilians with expertise in social media activities and planning, in-person interview with the authors, December 6, 2022.

[16] Civilians with expertise in social media activities and planning, in-person interview with the authors, December 6, 2022.

TABLE 2.1
Battle Rhythm Events and Key Attributes

Battle Rhythm Event	Key Attribute
IEWG	Planning event focused on layering multidomain effects across the 180–270-day time horizon. Uses published commander's guidance to focus OAI effects in line with the theater campaign order (TCO) and prioritized IMO.
JEWG	Planning event focused on resourcing, adjudicating requests, and deconflicting OAI across the 90–180-day time horizon. Also assesses the last 90-day OAI to act as inputs for future iterations.
SAWG	Planning event focused on coordinating special activities across the 180–270-day time frame.
JTWG	Long-standing WG that concentrates the delivery of effects on local activities and forces across the 180–270-day time horizon. It refines and publishes targeting guidance and tasks SCCs to prosecute targets.
PRAC	A biannual conference that aligns resources to OAI. It was instituted to help USCENTCOM leadership align with the PPB&E and GFMAP cycles.

Influence Effects Working Group

The IEWG's time horizon for planning is the longest, focusing on planning OAI three quarters (180–270 days) out from execution. The purpose of this WG is to identify long-term opportunities to exercise OAI within the IE using all capabilities to achieve the U.S. Central Command campaign plan (CCP) objectives across USCENTCOM LOEs.[17] It seeks to layer multidomain effects and capabilities from the SCCs and from USCENTCOM-retained capabilities. Capabilities designed to primarily cause effects in and through the IE are specifically considered by this WG. It focuses its planning activities on strategic guidance and the USCENTCOM commander's desired themes. While these themes can shift, they generally include SCC-focused OAI, building partnerships, and countering adversarial activities. They are also generally in line with the TCO and prioritize IMOs, which is how USCENTCOM measures progress toward its longer-term goals.

The IEWG meets biweekly and is chaired by the J39 Division Chief. It is a steady-state WG and includes participation from SCCs, coalition HQ, MISO, military deception (MILDEC), joint electromagnetic spectrum operations (JEMSO), OPSEC, STO, space, J5 assessments, WebOps, interagency coordination group, cyber, joint fires element (JFE), and intelligence personnel. External participants include Army 1st Information Operations Command, the JMWC, relevant TFs, and the Joint Staff's Joint Information Operations Warfare Center. Additional involved stakeholders can include Department of State personnel, partner-nation personnel, the Marine Corps Information Operations Center, individual country teams, and personnel from other GCCs.

[17] CENTCOM J39, "Joint Effects Working Group Charter," briefing, MacDill Air Force Base, September 7, 2021.

This WG is the first to consider layering effects within the command. There are multiple WGs that feed into the IEWG, which include the MEWG, OPSEC, JEMSO WG, Public Affairs WG, Key Leader Working Group, Commander's Synchronization Working Group (CSWG), and the Information Assurance WG. Outputs from these subordinate WGs are often used as inputs to the IEWG. Additional inputs include strategic guidance, influence-scoped intelligence preparation of the operational environment, commander's objectives, OAI, theater campaign plan (TCP) LOEs, operational assessments, and effects objective refinements. Outputs from the IEWG include recommended changes to the commanders' guidance and objectives, OAI initial concept sketches and problem statements, tasks to component staff, collection requirements and assessments, collection emphasis messages, and adversary-focused requests for information. Figure 2.3 highlights the IEWG information flow and shows many of the inputs and outputs from this battle rhythm event.

Insights into the Process

Through our research, we gained insight into this WG and its importance to the JEP. One insight came from an interviewee who talked about the tasking process related to the IEWG

FIGURE 2.3
Influence Effects Working Group Information Flow

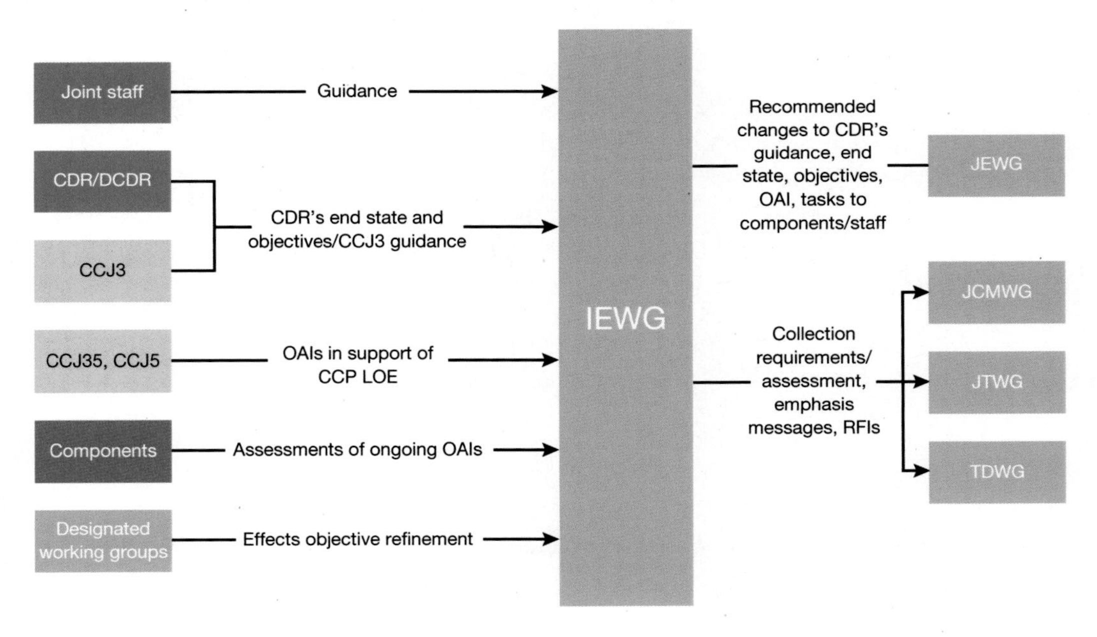

SOURCE: Adapted from USCENTCOM-provided briefing, September 7, 2021.

NOTE: CCJ3 = combatant command J3 (operations); CCJ35 = combatant command J35 (future plans); CCJ5 = combatant command J5 (plans); CCP = U.S. Central Command campaign plan; CDR = commander; DCDR = deputy commander; IEWG = influence effects working group; JCMWG = joint collection management working group; JEWG = joint effects working group; JTWG = joint targeting working group; LOE = line of effort; OAI = operations, activities, and investments; RFI = request for information; TDWG = target development working group.

and noted that the IEWG looks at the broader picture and works coordination, but direct tasking really does not come out of the process.[18] The IEWG uses the TCO as an input to the process, and, from it, IMOs are identified that lead to desired priority effects. Ideally, desired effects are derived from the IMOs in the plan, and the commander's guidance helps the staff plan OAI to achieve them. However, the components control most of the forces and resources to conduct OAI. This creates a situation in which the IEWG identifies priority effects and then should publish an order to help align information efforts. In reality, this order is not routinely published, and there is no forcing function to make the SCCs conduct specific OAIs.[19]

Another insight pertains to the layering of effects. As mentioned previously, one of the main purposes of the IEWG is to identify potential effects and assist in developing OAI to achieve them. As one interviewee noted, there are times when CONOPS received from components are missing basic details.[20] For example, in an operation that needed to leverage cyber effects, a cyber element already had a product developed and wanted to use it. The J39 had to work backward from this product to develop a CONOPS so that it could be properly staffed and then pushed forward to the JEWG. The interviewee noted that this is contrary to the intended process but occurs frequently.[21] A way to mitigate this could be to connect the cyber element with one of the subordinate WGs (in this instance, the MEWG was suggested) before the IEWG meets. This would help to align both the cyber and MISO effects at a lower-level WG so that when they present the concept at the IEWG, it is a more robust product for discussion. The IEWG could then staff the CONOPS for future action and consideration at the JEWG.

Other issues with the IEWG relate to the WGs that feed into the IEWG. One of these is the key leader engagement working group (KLEWG), which is run by the J5. As one interviewee noted, the J5 needs to refine the process that integrates key leader engagement (KLE) objectives with CUOPS and desired effects.[22] This individual noted that the KLEWG needs to get beyond "the product" to "the purpose." While a spreadsheet for engagements should be kept and updated, the USCENTCOM commander's front office plans a host of engagements, which are not often synched with the JEP. One way to improve this process would

[18] Military officers with expertise in planning and operations, in-person interview with the authors, September 14, 2022.

[19] Military officer with expertise in planning and operations, in-person interview with the authors, December 5, 2022.

[20] Military officer and civilian with expertise in psychological operations (PSYOP) and planning, in-person interview with the authors, December 6, 2022.

[21] Military officer and civilian with expertise in PSYOP and planning, in-person interview with the authors, December 6, 2022.

[22] Military officers with expertise in planning and operations, in-person interview with the authors, September 14, 2022.

be to incorporate key USCENTCOM leaders' calendars into the IEWG.[23] Travel dates and engagements could be categorized as OAI then vetted in the same way as other proposed CONOPS. An output from the KLEWG could then be used as an input to consider for future IEWGs. This would allow USCENTCOM to, as some interviewees described it, "weaponize the calendar."[24] Some potential constraints to this improvement include the need to identify authors and a process for producing relevant information papers before KLEs, overcoming the close-hold nature of the commander's calendar for OPSEC concerns and increasing distribution of lists of the readouts from these events, as action officers within the various J-code directorates receive this information only if someone forwards it.[25]

A final consideration concerns the general understanding of influence and what that means. As one interviewee noted, "there needs to be more people that understand influence to get a handle on the situation."[26] One venue to help improve understanding could be the MEWG, which was previously a defunct WG but was brought back as part of a staff reorganization and reallocation in October 2021. Although the MEWG is now back in operation, there is still a lack of influence professionals in the command. For example, there are only a limited number of PSYOP personnel in the J39, which are split between FUOPS and CUOPS. Down at the SCCs and other subordinate elements, there may only be one PSYOP-trained person, with some elements having none.[27]

Joint Effects Working Group

The purposes of the JEWG are to develop and refine the TCO, IMO, and desired effects and to draft guidance to integrate capabilities for targeting opportunities.[28] It operates between 90 days and 180 days forward in the planning cycle and supports the JEB by managing resources through the execution of an OAI. It meets weekly and is chaired by the J39 director of operations. It is both a steady-state and crisis-action meeting. Internal participants include the JFE; representatives from multiple J3 directorates; intelligence, plans, and security cooperation; staff judge advocates; and cyber elements. External representatives include component repre-

---

[23] Military officer with expertise in OIE and planning, in-person interview with the authors, September 15, 2022.

[24] Military officers with expertise in planning and operations, in-person interview with the authors, September 14, 2022.

[25] Military officer with expertise in OIE and planning, in-person interview with the authors, September 15, 2022.

[26] Military officer and civilian with expertise in PSYOP and planning, in-person interview with the authors, December 6, 2022.

[27] Military officer and civilian with expertise in PSYOP and planning, in-person interview with the authors, December 6, 2022.

[28] CENTCOM J39, 2022.

sentatives and TFs elements. Additional members could include the U.S. Department of State, U.S. Agency for International Development, and J6 and J8 personnel.

Inputs into the process include strategic guidance and objectives; OAI; desired OAI effects; post-OAI assessments; desired effects and prioritization; last 90-day assessments; intelligence, surveillance, and reconnaissance priorities and plans; IMO and desired effects; and operational assessments. Outputs from the JEWG are draft effects objectives and desired effects, tasks to subordinates, recommended changes to the CCP LOEs, operational effects assessments, component and staff OAI (previous 90 days through next 180 days), adversary opportunities to exploit, and recommended changes to ISR priorities and plans. Figure 2.4

FIGURE 2.4
Joint Effects Working Group Information Flow

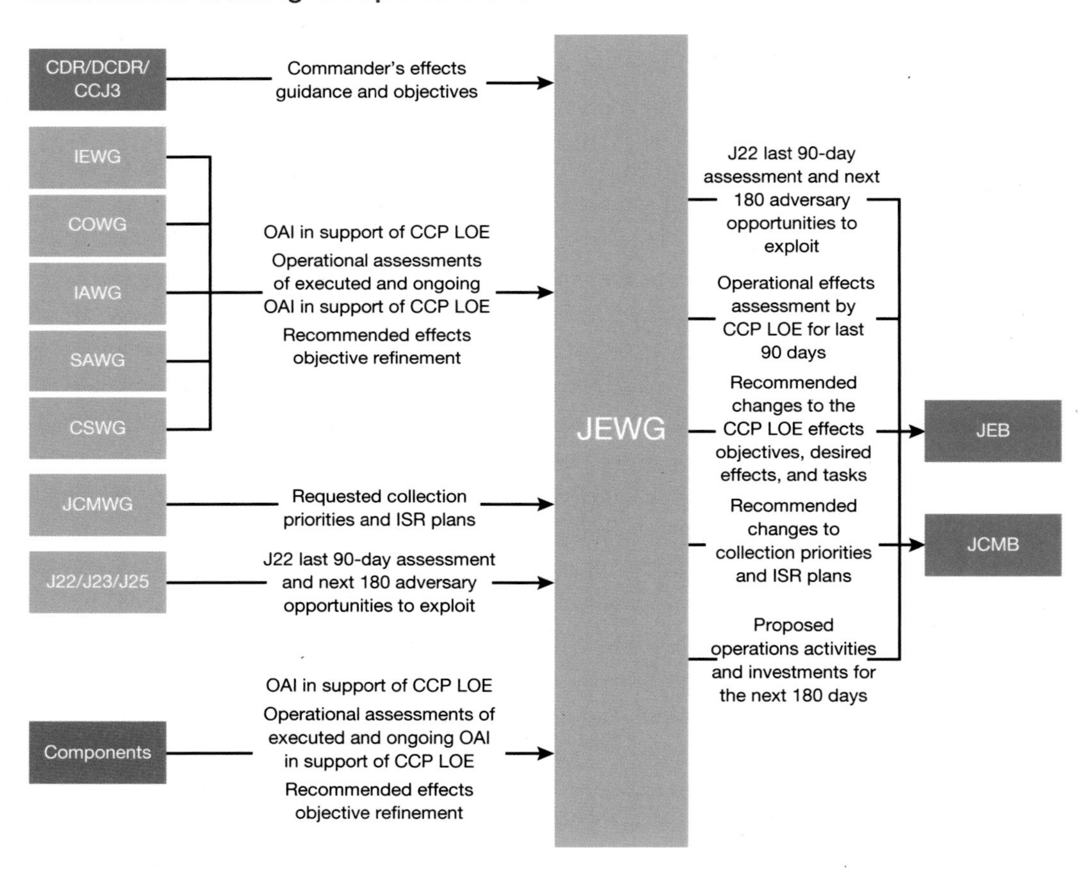

SOURCE: Adapted from USCENTCOM-provided briefing, March 15, 2022.

NOTE: CCJ3 = combatant command J3 (operations); CCP = U.S. Central Command campaign plan; CDR = commander; COWG = cyber operations working group—replaced with cyber effects working group; CSWG = Commander's Synchronization Working Group; DCDR = deputy commander; IAWG = inter-agency working group; IEWG = influence effects working group; ISR = intelligence, surveillance, and reconnaissance; J22 (future plans); J23 (CUOPS); J25 (future plans); JCMB = joint collection management board; JCMWG = joint collection management working group; JEB = joint effects board; JEWG = joint effects working group; LOE = line of effort; OAI = operations, activities, and investments; SAWG = special activities working group.

shows the JEWG information flow, supporting WGs, inputs and outputs from this WG, and how they connect to the larger effects process.

Insights into the Process

The JEWG was created in 2021 to bring together disparate activities being conducted by the SCC and across the staff and to help focus effects and planning. The JFE was initially tasked to create the process because there is a natural fit with artillery personnel, targeting methodologies, and measuring effects. While the targeting process is effective for the lethal prosecution of targets, it is not as conducive to operations that need more time to generate effects or directs effects at entities other than targets.[29] As generating influence in multiple target audiences often requires substantial amounts of time, the JFE-led process did not adequately address these longer time horizons. This is typical for most behavioral changes, which usually require sustained influence campaigns.[30] This ultimately led to a shift in directorates that chaired the JEWG. In September 2021, the J39 took over the JEWG from JFE, and by the spring 2022 the JEWG had significantly evolved.[31] As the process evolved, several insights into the new process emerged.

As the process evolved, the organizing principle for the WG changed several times. First, it changed focus from LOEs to more-discrete regional areas; subsequently, it changed to focus on IMOs. While some SCC representatives wanted to continue to orient around LOEs or regional alignments, other stakeholders did not like these orientations. Orientation around IMOs sought to group similar activities with similar purposes. With IMO binning, the process helped drive discussions toward capacity building and helped to overcome some of the challenges related to layering effects. Interviewees noted that when the process focused only on LOEs, there was rarely overlap in OAI, and it was difficult to track how capacity-building efforts advanced progress on LOEs. Instead, organizing around and focusing on IMOs helps to refine assessments because there are more discrete activities to measure, thereby making

[29] Per an in-person discussion between the authors and Marine Corps doctrine writers and personnel with experience in the Marine Corps terminology office on March 30, 2023, by law and policy, *target* refers only to enemy or adversary systems or personnel that are valid subjects of military force; entity is a much more inclusive term for people or systems who might be subject to effects (whether through targeting or through some other process). Terms other than entity and target are also possible. Joint Publication 3-04 refers inclusively to "relevant actors," MISO address "target audiences," and Joint Publication 3-61 notes that public affairs concerns itself with audiences, publics, and stakeholders. *Entity* is the broadest appropriate term for persons, groups, or systems that might be subject to some sort of intentional effect from OAI. See glossaries in Joint Publication 3-04, 2022; and Joint Publication 3-61, *Public Affairs*, U.S. Joint Chiefs of Staff, incorporating change 1, August 19, 2016.

[30] Retired senior military officer with expertise in OIE, planning, and operations, telephone interview with the authors, November 7, 2022.

[31] Military officer with expertise in planning and operations, in-person interview with the authors, December 5, 2022.

a stronger logical link between OAI and desired effects.[32] By concentrating on IMO effects, planners can consider a timeline, geographic boundaries, and effects simultaneously, which has helped to ease the confusion. However, interview subjects asserted that the IMOs are still not sufficiently precise. As one interviewee noted, the IMOs need to be more specific and measurable; as currently written, they do not provide actionable steps.[33] The interviewee went on to say that the flaw lies with the campaign plan itself and that the process to generate effects is only as good as the IMOs they support. They asserted that a proper solution would involve writing truly attainable IMOs into the TCP.[34]

The JEWG has also significantly improved the planning processes of subordinate SCCs. For example, the Marine Corps Forces Central Command commander directed the creation of a new campaign plan that provides specific guidance for priority OAI along with the rationale as to why they are priorities. In this campaign plan, OAI are tiered into multiple categories and then rated. This was done because the JEWG forced the SCCs to brief their top three OAI at each meeting. According to one individual, this has worked well as a forcing function.[35] Marine Corps Forces Central Command uses these prioritized OAI to develop its concepts of support to pitch at the JEWG. The use of command and control in the information environment (C2IE), which is the program of record software tool that USCENTCOM uses to track all OAI, has added further fidelity and insight into these OAI.[36]

Additionally, as the structure of the JEWG evolved, it has brought new J-codes and action officers into the process. For example, intelligence professionals now start every meeting with an intelligence update. In the words of one interviewee, this structure has provided an opportunity for intelligence professionals to "speak truth to power."[37] JEWG sessions begin with an adversary focus on both activities and motivations rather than activities only.[38]

A final insight into the evolution of the JEWG occurred during a staff planning exercise that examined many staff processes, including the JEWG. The USCENTCOM staff looked at the last 180 days and conducted a self-evaluation. One interviewee noted that this evaluation might have had the pitfall of many self-assessments, which they viewed as "often too

[32] Senior military officers with expertise in OIE and planning, in-person interview with the authors, December 5, 2022.

[33] Military officer with expertise in planning and operations, in-person interview with the authors, December 5, 2022.

[34] Military officer with expertise in planning and operations, in-person interview with the authors, December 5, 2022.

[35] Military officers with expertise in planning, in-person interview with the authors, December 6, 2022.

[36] Senior military officers with expertise in OIE and planning, in-person interview with the authors, December 5, 2022.

[37] Senior military officers with expertise in OIE and planning, in-person interview with the authors, December 5, 2022.

[38] Senior military officers with expertise in OIE and planning, in-person interview with the authors, December 5, 2022.

positive."[39] To rectify this perception, USCENTCOM has shifted the paradigm to be more critical of its assessments and to better consider future OAI based on more-robust assessments provided during the JEWG. Specifically, USCENTCOM leadership wants to see how changes to the TCP can help lead to better linkages between planning OAI and achieving measurable effects.

Joint Effects Board

JEB activities are the culminating events for the IEWG and JEWG. The board's purpose is to approve OAI, integrate effects across time horizons, review and assess operational effects,

FIGURE 2.5
Joint Effects Board Information Flow

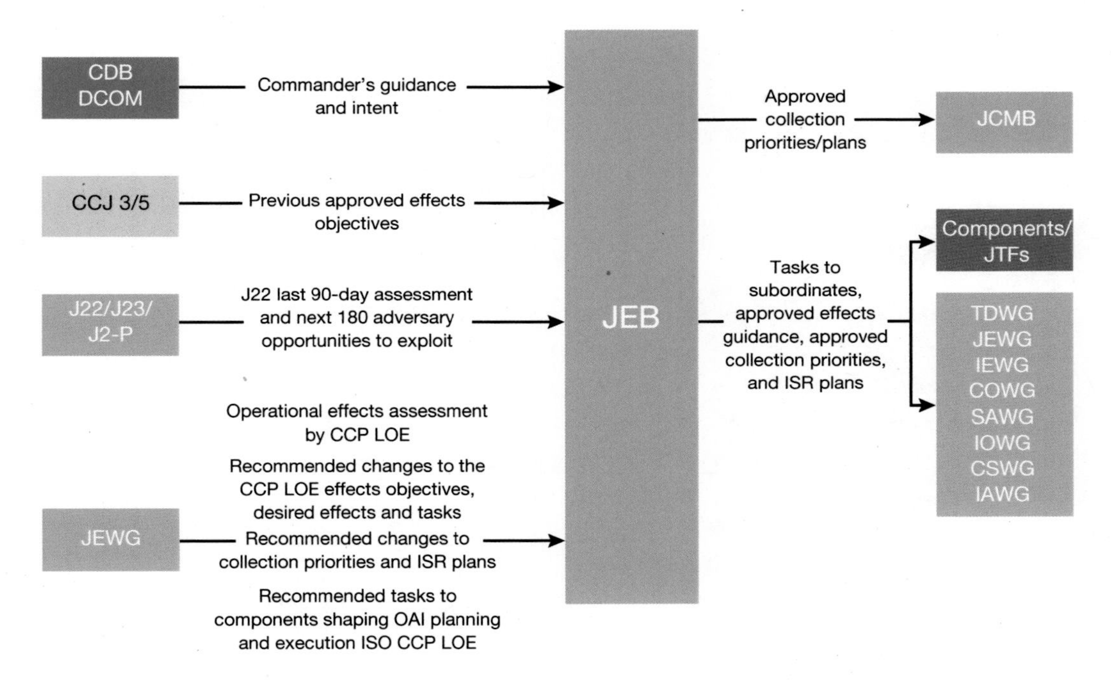

SOURCE: Adapted from USCENTCOM provided briefing, March 15, 2021.

NOTE: CCJ3 = combatant command J3 (operations); CCJ5 = combatant command J5 (plans); CCP = U.S. Central Command campaign plan; CDB = Command Decision Board; COWG = replaced with cyber effects working group; CSWG = Commander's Synchronization Working Group; DCOM = deputy commander; IAWG = inter-agency working group; IEWG = influence effects working group; IOWG = information operations working group—defunct; ISR = intelligence, surveillance, and reconnaissance; J22 = Future Plans; J23 = CUOPS; J25 = Future Plans; JCMB = joint collection management board; JEWG = joint effects working group; JTF = joint task force; LOE = line of effort; SAWG = special activities working group; TDWG = target development working group.

[39] Senior military officers with expertise in OIE and planning, in-person interview with the authors, December 5, 2022.

allocate resources to SCC requirements and requests, and assess the effectiveness of current targeting efforts. It operates on a six-week cycle and approves activities approximately 90 days in advance of execution. The deputy director of operations chairs the board during both steady-state and crisis-action horizons. Participants include the J39, JFE, and action officer-level representatives from intelligence (J2) plans (J5), Staff Judge Advocate, and senior-level SCC, JTF, and TF personnel. Inputs into the board include commanders' guidance and intent, previous approved objectives, operational assessments of the CCP by LOE, recommended changes to LOE effects, objectives, and tasks, the J2 90–180-day assessment of adversary activities, and recommended tasks for SCC OAI. Board outputs include approved collection priorities and plans, tasks to subordinates, SCC and staff OAI for the next 180 days, approved changes to effects objectives, and approved ISR plans and priorities. Figure 2.5 highlights the JEB information flow and inputs and outputs, as well as shows how the JEB closes the JEP.

Insights on the Process

The JEB is the approval venue for all the work conducted in the IEWG and JEWG. It also considers OAI proposed through the SAWG. The JEB meets once every six weeks and is supposed to produce an effects order. As of late 2022, however, these effects orders have not been produced, and no other interviewees mentioned or provided an example of such an order.[40] However, the JEB has significantly helped to codify the entire process into the USCENTCOM battle rhythm. A comparison with the previous process highlights how these changes have occurred.

For example, the architecture for this previous process, which was used before the new JEP and its attendant specific B2C2WGs were adopted, was contained in several PowerPoint slide decks that had limited information and showed the location of OAI on a map. Updating slides was very time intensive, and the SCCs did not understand all the OAI that were occurring. During this stage, the insights from the process were narrowly focused and not cross-cutting functional areas, SCCs, or the staff. Furthermore, they provided little additional understanding of USCENTCOM activities.[41] Some of the SCCs were not even tracking all of the OAI that they were conducting.[42] As one interviewee noted, USCENTCOM operated on a six-week cycle, could not plan more than two quarters out (thus limiting its ability to participate in global exercises and events), and could not align resources effectively.[43] The JEB was the culmination of this process but was not centralized. Instead, the JEB was conducted via regular briefs, sometimes at the same meeting and sometimes at separate meetings with

---

[40] Military officer with expertise in planning and operations, in-person interview with the authors, December 5, 2022.

[41] Military officers with expertise in planning, in-person interview with the authors, December 6, 2022.

[42] Military officers with expertise in planning, in-person interview with the authors, December 6, 2022.

[43] Military officer with expertise in planning and operations, in-person interview with the authors, December 5, 2022.

various J-code directorates. After these took place, a final briefing was given to the deputy commander approximately every six weeks. This six-week cycle was focused on effects-based planning through a weekly WG that focused on one LOE and was bounded by geographic areas.[44] The staff would weave together different activity, operations, and effects assessments from the J3 and J5. As one interviewee noted, "[I]t was not well coordinated . . . a lot of people were burning JP-8 [airplane fuel] to make noise, but not much of substance was produced."[45]

Being chaired by the deputy director for operations, a three-star general or flag officer, lends weight to the process. Tasking from the JEB codifies planning efforts across B2C2WG, the staff, SCCs, and JTFs. It directs collection assets to align with planning efforts and upcoming OAI. Whereas before the entire JEP was disjoined, the JEB has codified a process that incorporates influence and influence effects into mainstream USCENTCOM planning.

Special Activities Working Group

The purpose of the SAWG is to review, coordinate, deconflict, synchronize, and measure progress of proposed operations under a special access program, STO, Noncombatant Evacuation Order Working Group, and MILDEC. It is a weekly WG that is chaired by the J39 director. It is both a crisis-action and steady-state WG. Internal members include component HQ, intelligence, operations, cyber, and staff judge advocates. External participants include the SCCs, TFs, U.S. Special Operations Command (USSOCOM), National Security Agency, and other intelligence community special advisors. Other members can be brought into the WG as needed.[46]

Product inputs for this WG include the commander's objectives, strategic guidance, SCC and other command requirements, and all directorate and SCC special action program, STO, and other Alternative or Compensatory Control Measures operations. Outputs from this WG include highly sensitive CONOPS that could not be discussed in the other B2C2WG, operational updates, and recommendations for sensitive and special activities and coordination with other members and other governmental agencies, as well as planning guidance for FUOPS. Figure 2.6 shows the SAWG information flow.

Insights on the Process

The SAWG is similar to the IEWG in that it tries to synchronize and layer effects across a long-range time horizon. Both these WGs are about creating effects and synergy across the

[44] Senior military officers with expertise in OIE and planning, in-person interview with the authors, December 5, 2022.

[45] Retired senior military officer with expertise in OIE, planning, and operations, telephone interview with the authors, November 7, 2022.

[46] Military officer with expertise in operations and planning, in-person interview with the authors, September 15, 2022.

FIGURE 2.6
Special Activities Working Group Information Flow

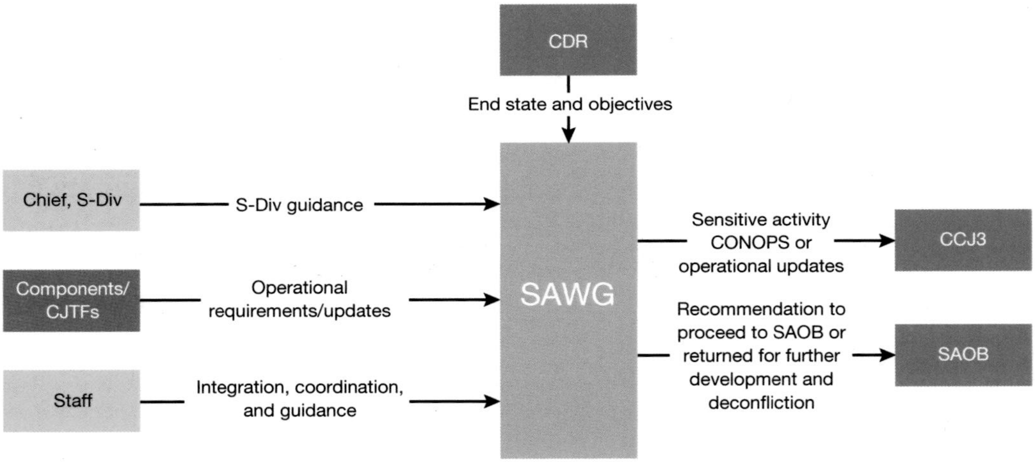

CDR
End state and objectives
Chief, S-Div
S-Div guidance
Components/ CJTFs
Operational requirements/updates
Staff
Integration, coordination, and guidance
SAWG
Sensitive activity CONOPS or operational updates
CCJ3
Recommendation to proceed to SAOB or returned for further development and deconfliction
SAOB

SOURCE: Adapted from USCENTCOM provided briefing, April 22, 2021.
NOTE: CCJ3 = combatant commander J3 (operations); CDR = commander; CJTF = Combined Joint Task Force; CONOPS = concept of operations; SAOB = Special Activities Oversight Board; SAWG = special activities working group; S-Div = Special Activities Division.

staff, SCCs, and capability areas.[47] Much like the IEWG and JEWG, the SAWG has evolved substantially since 2021, after having gone defunct for some time. Prior to its rejuvenation, the SAWG was an existing WG that was run by J3 personnel who were primarily concerned with having the SCCs brief them on the OAI they were conducting.[48] As the IEWG and JEWG began to change form, the SAWG did too. For example, as the SAWG evolved, the USCENTCOM staff started to take more of a leading role, establish a more formal framework for planning operations, and compel the SCCs to plan and execute within that framework. While the 20 or so personnel who attended it remained relatively steady throughout its transformation, the intent and mechanisms changed.[49]

The largest change focused on the intent of the SAWG. Whereas previously, the SAWG received briefs from the SCCs on what they were doing, there was relatively little layering of effects or direction provided from USCENTCOM. The SAWG sought to change this by ensuring that USCENTCOM drove the effects process to reduce the likelihood of possibly contradictory effects across OAI and confusion across SCC and staff. In much the same way

[47] Military officer with expertise in planning and operations, in-person interview with the authors, December 5, 2022.

[48] Retired senior military officer with expertise in OIE, planning, and operations, telephone interview with the authors, October 5, 2022.

[49] Retired senior military officer with expertise in OIE, planning, and operations, telephone interview with the authors, October 5, 2022.

that the IEWG started to push forward CONOPS from lower-tier WGs and SCCs, the SAWG adopted the same framework. The SAWG also began to focus on assessments and the use of intelligence to help drive OAI,[50] which was further helped by several key leaders who contributed to both WGs, leading to cross-pollination between the SAWG and the JEWG.

USCENTCOM needed to do more than just take and pass information through the command; it needed to deconflict operations, add resources and additional capabilities, and approve sensitive activities. To help achieve this, USCENTCOM created the SAOB (chaired by the director of operations [J3]) and Special Activities Oversight Council (SAOC) (chaired by the USCENTCOM commander). The SAOB was on the same relative six-week cycle as the other influence-focused WGs, and the SAOC met quarterly and was aligned to the Combatant Commander Conference.

As one interviewee noted, the SAOB and SAOC pulled together multiple OAI at one time and in one meeting to help make sense of all special activities for the USCENTCOM commander and other subordinate commanders, which something that was always difficult to achieve because of the highly classified nature of these activities.[51] At both the SAOB and the SAOC, a more stringent level of oversight was applied.

Joint Targeting Working Group

The JTWG is one of the most well-established events at USCENTCOM and is chaired by the joint fires element. One level up, the deputy director of operations chairs the Targeting Validation Board. Above that, the USCENTCOM commander chairs the Joint Targeting Coordination Board. The JTWG meets weekly and spends a portion of its time refining targeting guidance and tasking SCCs. The Joint Targeting Coordination Board produces refined targeting guidance and validates targets but does not focus on tasking SCCs to develop or prosecute targets. As the processes for the JEWG evolved, USCENTCOM noted that there was still a need for a JTWG that focused its efforts on target development, validation, and tactical activities. Many of the products used in the JTWG, including adversarial assessments, were still effective and helpful to other B2C2WG, including the IEWG and JEWG. These products allow for a narrow focus on adversary forces and capabilities, which can help to bring a level of specificity to other WGs that are not developing those types of products.[52] Figure 2.7 shows the JTWG information flow.

---

[50] Military officer with expertise in operations and planning, in-person interview with the authors, September 15, 2022.

[51] Retired senior military officer with expertise in OIE, planning, and operations, telephone interview with the authors, November 7, 2022.

[52] Senior military officer with expertise in fires and planning, in-person interview with the authors, December 5, 2022.

FIGURE 2.7
Joint Targeting Working Group Information Flow

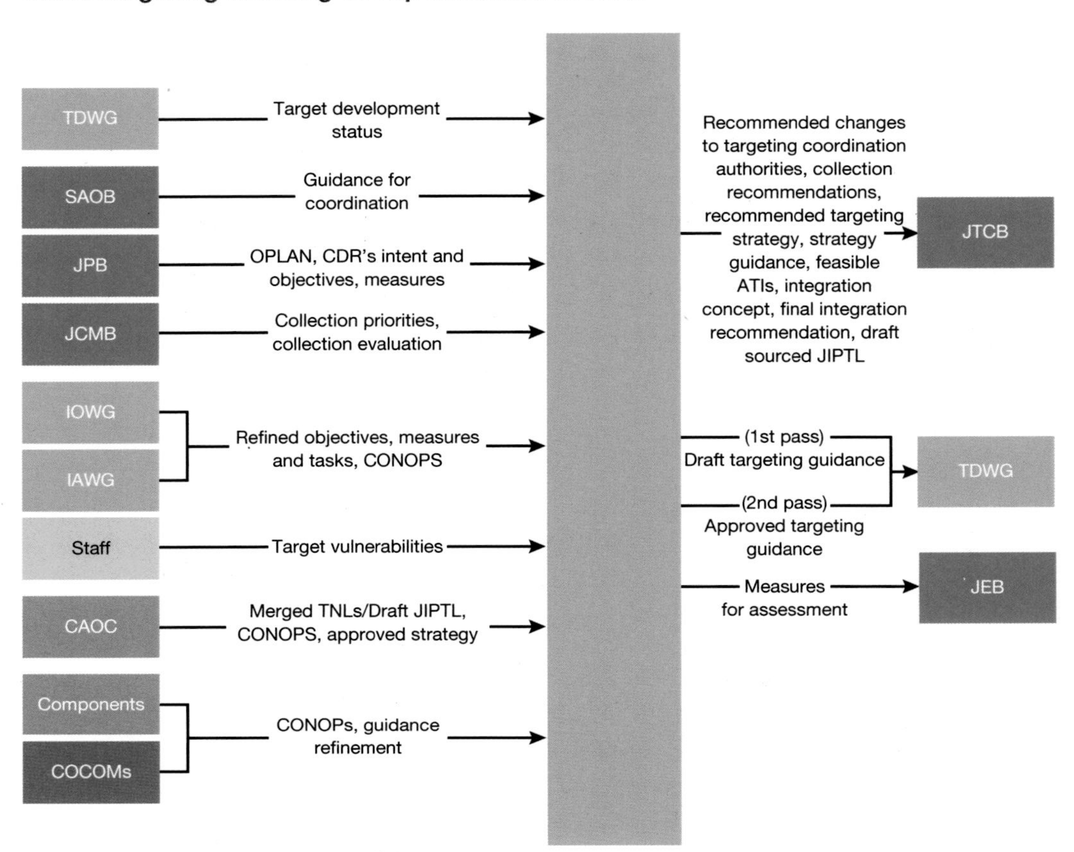

SOURCE: Adapted from USCENTCOM-provided briefing, December 1, 2021.
NOTE: ATI = asset target interactions; CAOC = Combined Air Operations Center; CDR = commander; COCOM = combatant command; CONOPS = concept of operations; IAWG = inter-agency working group; IOWG = information operations working group—defunct; JCMB = Joint Collection Management Board; JEB = joint effects process; JIPTL = joint integrated prioritized target list; JPB = Joint Planning Board; JTCB = joint target coordination board; OPLAN = operation plan; SAOB = Special Activities Oversight Board; TNL = target nomination list; TDWG = target development working group.

Insights on the Process

The main difference between the IEWG and JEWG and the JTWG is their focus. While the IEWG and JEWG focus on achieving effects that advance the campaign plan, the JTWG is adversary-focused.[53] In other words, the JTWG concentrates its efforts on local activities and forces and tactical fighting, which is different from the effects or campaign focus of the JEWG. The differences can be observed in how information flows through these two differ-

[53] Senior military officer with expertise in fires and planning, in-person interview with the authors, December 5, 2022.

ent processes and the time horizons that each covers. Because the JTWG is tactically focused, the time and place to engage a target tend to be relatively fixed. To support this process, the joint force uses a detailed database to hold the data for targets. Additionally, there is a well exercised targeting methodology to support the process, as well as a standardized periodic review process, which usually takes place every two years.[54]

However, nonkinetic and influence-focused target analysts must continually update their analyses in C2IE, a program which itself continues to evolve to meet emerging requirements to plan and capture cognitive-focused content. While C2IE is good at showing OAI as well as allowing access to assessment data, the targeting process does not use C2IE, nor does the JFE. Between the JEWG and JTWG, all effects are supposed to be integrated, and, ideally, this would be the case. As these processes are evolving, it was unclear to the research team whether the command is currently achieving its desired level of integration among the JTWG and the rest of the JEP. According to one perspective, the nearer-term tactical focus of the JTWG and the broader campaign focus of the JEWG created somewhat of a "mismatch" in synergy between the two.[55]

Planning and Resource Alignment Conference

To help align the three processes' (1) influence and effects, (2) special activities, and (3) targeting, USCENTCOM implemented the PRAC in 2021. The PRAC is specifically intended to help align resources to OAI.[56] The PRAC was loosely based on the USSOCOM global synchronization model, which aligns and resources multiple USSOCOM SCC activities. However, the USCENTCOM process differs slightly because of service priorities, budget processes, and PPB&E. The PRAC operates over a three-year planning horizon in which the first two years are used to plan, and the final year is used to confirm and then executes OAI. One interviewee noted that in a best-case scenario, the first year is spent planning, the second year is spent resourcing, and the third year is for execution.[57]

The first PRAC was held in April 2021. It started with a knowledge process flow map in which participants helped to align upcoming OAI. Representatives from the SCCs attended, briefing attendees about their upcoming OAI and the reason they were conducting these OAI. As one interviewee noted, this first iteration of the PRAC was not very smooth, and the information flow from the SCCs to USCENTCOM and vice versa was disjointed.[58] The

---

[54] Senior military officer with expertise in fires and planning, in-person interview with the authors, December 5, 2022.

[55] Senior military officer with expertise in fires and planning, in-person interview with the authors, December 5, 2022.

[56] Military officers with expertise in planning, in-person interview with the authors, December 5, 2022.

[57] Military officers with expertise in planning, in-person interview with the authors, December 5, 2022.

[58] Military officers with expertise in planning, in-person interview with the authors, December 5, 2022.

PRAC continued to evolve and improve over subsequent iterations. By March 2022, the format included each SCC briefing their OAI and the attendees ranking them in terms of which would achieve the most important effect. The key factors weighed in the voting include the scope and scale of specific OAI, internal SCC priority level, and the location of each of the OAI.[59] Campaign assessment was incorporated into the process by September 2022 using data from the USCENTCOM GFMAP process and various J8 assessments.[60]

Insights on the Process

As USCENTCOM evolved its processes, it recognized that resources needed to be aligned with OAI on a longer timescale to account for layer effects and so that the SCCs could take into consideration all the things that they were doing. Senior leaders acknowledged that aligning USCENTCOM processes to the PPB&E and GFMAP cycles would present "windows of opportunity" for planners.[61] However, this requires a six-month cycle to allow the SCCs and staff to get fully organized. To describe this challenge, one interviewee noted that it was decided to send a Carrier TF into the theater during the initial PRAC. However, there was no discussion about why the command was doing this or what effect USCENTCOM wanted to achieve. Because the command had access to a Carrier TF, it was able to send it into theater without considering the influence effects.[62] USCENTCOM needed to flip the planning paradigm to first identify the effect to be achieved, align OAI to achieve it, then resource those OAI. The PRAC filled the resourcing gap of this process.

As discussed, the PRAC has evolved to identify priorities and conduct some level of assessment with greater rigor, but there are still significant challenges, specifically when it comes to risk articulation. The command does not currently have consensus on how to characterize and discuss risk in the IE; also, it does not formally articulate the risk of inaction. For example, the PRAC does not articulate the risk to the campaign plan if a specific OAI is not conducted. While USCENTCOM does have a Joint Risk Assessment Matrix and other risk management frameworks, these frameworks do not sufficiently address IE issues. Relatedly, the PRAC helped identify and improve shortcomings in the C2IE system, such as resourcing and prioritization. One gap is that C2IE still lacks a mechanism for communicating risk.[63]

After its steady evolution, the PRAC has significantly helped to evolve USCENTCOM planning efforts, and there are many problems that the PRAC has helped to overcome. It helped to align and resource SCC OAI, bring assessment into the process, and focus activities to align with the CCP. Additionally, the PRAC helped fix two problems: (1) USCENTCOM

[59] Military officers with expertise in planning, in-person interview with the authors, December 5, 2022.

[60] Military officer with expertise in assessment, in-person interview with the authors, September 16, 2022.

[61] Military officers with expertise in planning, in-person interview with the authors, December 5, 2022.

[62] Military officers with expertise in planning, in-person interview with the authors, December 5, 2022.

[63] Retired senior military officer with expertise in OIE, planning, and operations, telephone interview with the authors, November 7, 2022.

did not know what it wanted to do versus what it needed to do, and (2) USCENTCOM previously thought in terms of capabilities, not in terms of the effects it wanted to achieve. The PRAC helped to address both these issues.[64] Finally, the establishment of the PRAC met the USCENTCOM commander's intent to drive positive discussion and provide more-refined campaign guidance so that the SCCs could plan and execute OAI with the help of resources from the PRAC.

Conclusion

The implementation of and reforms to the IEWG, SAWG, JEWG, PRAC and JTWG comprise a unique process for joint effects. This JEP is substantively different from USCENTCOM's previous battle rhythms in terms of its format and its substance. It seeks to put effects-based operations into practice, to elevate the role of information in planning, and to make it central to everything that the GCC does. It was clear from the SMEs interviewed by RAND that the JEP is still a work in progress. In some areas, it has made the battle rhythm better (that is, more effective at integrating information activities), while in others, there is still room for progress. In Chapter 4, we leverage the literature on military operations, the experience of the RAND team, and the feedback from the SMEs engaged in this project to develop a framework for assessing a battle rhythm. This framework provides a guide for thinking about what makes a battle rhythm more or less effective and whether and how the JEP serves its intended purpose.

---

[64] Military officers with expertise in planning, in-person interview with the authors, December 5, 2022.

CHAPTER 3

Assessing Process: What Characteristics Are Desirable in a Joint Effects Process?

The intended aim of the revised USCENTCOM JEP is to better integrate considerations and planning related to the information joint function and OIE.[1] Interviews, document reviews, and case studies of selected operations revealed that the USCENTCOM JEP has largely met that aim. However, documenting successes is only a part of the analytic objective of this research. We also sought to understand what characteristics and features of the JEP led to its success so that those characteristics and features can be preserved, further improved on, or replicated in the future. This chapter presents a framework for assessing a joint effects–type process on the features and characteristics that support the effective integration, inclusion, and emphasis of the information joint function.

Framework for Assessing the Inclusion of Information in a Joint Effects Process

The framework distills and enables the application of insights gained through this and other research efforts. The framework builds on our previous experience with OIE-related staffing processes and arrangements, as well as insights gleaned from the interviews, command documents, and case studies. The core of the framework is a list of desirable characteristics and features (in other words, staff processes that are designed to achieve or sustain characteristics or features that improve or sustain the inclusion and integration of OIE and information as a joint function in the command's staff processes). To populate the list, we sought characteristics and features that were clearly connected to intended outcomes or were flagged as important contributors to process effectiveness by interview participants. We also sought to identify negative characteristics and features that were altered or intentionally abandoned from previous versions of the command's battle rhythm events. Working independently and

[1] Information was formally added to the list of joint warfighting functions in 2017. For an overview on how this change reflects an evolution of joint doctrine, see Keith Adkins and Tom Evans, "Information as a Joint Function: A Doctrinal Perspective," briefing slides, Joint Information Operations Proponent, Joint Staff J39 Strategic Effects Division, Deputy Director for Global Operations, May 4, 2018.

using these resources, several authors compiled candidate lists of process characteristics favorable to desired outcomes. These were then reconciled into a single list, which was validated through a presentation to stakeholders who were previously interviewed as part of the research.

Table 3.1 lists the desirable features and characteristics. These 19 characteristics support four major thematic categories: An effective JEP

- (C) centralizes and elevates information power and related effects
- (A) aligns activity and strategy
- (I) integrates, synchronizes, synergizes, and layers informational and physical power
- (F) provides feedback and promotes iterative improvement.

This list becomes an evaluative framework when used as criteria to assess an existing or proposed collection of staff processes and their likelihood of fostering the effective integration of information with other command OAI. In the remainder of the chapter, we discuss the virtues of each of the listed features. (See Chapter 4 for an application of the framework.)

Category C: Centralizes and Elevates Information Power and Related Effects

This thematic category includes four desirable characteristics or features: (C1) receives significant attention from senior leadership, (C2) centralizes and elevates planning rather than leaving it at the service components, (C3) elevates the prominence of information in command processes and in SCC awareness and broadens staff awareness across the range of kinetic and nonkinetic OAI, and (C4) involves the right people in the right numbers.

C1. Receives Attention from Senior Leadership

Effective B2C2WG must involve senior leaders as an endorsement of its purpose. By *senior leaders,* we mean flag or general officer-level directors or chiefs of the command's staff structure (the so-called "J-Dirs") or the commander or deputy commander. To effectively integrate the information joint function in a GCC effects process, senior leaders must be seen as engaged in some elements of the process and interested in the outputs of the process (as support to their decisionmaking). Senior leaders signal interest by attending boards or consolidation battle rhythm events in person and by making sure representatives from their directorate attend supporting WGs. Senior leaders show interest in outputs by incorporating them in plans, endorsing them, using them to support their decisionmaking, and giving feedback and acknowledgment. The demonstrated interest and investment by senior leaders can motivate and incentivize subordinate staffs to allocate time and effort to meetings and processes. When leaders actively use B2C2WG, subordinates see the events as relevant to accomplishing the mission. Similarly, their attendance can demonstrate leaders' expectations their staff to

TABLE 3.1
Characteristics and Features for Assessing the Inclusion of Information in a Joint Effects Process

Category	Characteristic or Feature
Centralizes and elevates information power and related effects	
C1	Receives significant senior leader attention
C2	Centralizes and elevates planning rather than leaving it at the service components
C3	Elevates prominence of information in command processes and in SCC awareness; broadens staff awareness across the range of kinetic and nonkinetic OAI
C4	Involves the right people in the right numbers
Aligns activity and strategy	
A1	Promotes a campaigning mindset
A2	Aligns and links OAI with command objectives
A3	Generates unified, synchronized, justified, and well-timed force planning requirements
A4	Demands justification of habitual and routinized events; takes compulsory events and layers additional OAI to support them in ways that better tie them to theater campaign and IMOs
A5	Supports or bridges a medium-scale planning time horizon (between CUOPS and FUOPS)
Integrates, synchronizes, synergizes, and layers informational and physical power	
I1	Is able to generate CONOPS that support multiple objectives/levels; able to coordinate multiple OAI against a single effect or objective
I2	Deconflicts and coordinates authorities and permissions across the service components and other elements
I3	Promotes/enables information sharing across relevant staff sections, components, and capabilities
I4	Integrated with intelligence such that the process is intelligence driven and encourages additional intelligence support
I5	Accounts for and integrates special activities
Provides feedback and promotes iterative improvement	
F1	Is able to provide effective feedback and input to the TCP and TCO process and promote clear IMOs
F2	Supports dynamic/evolving objectives between iterations of TCO cycle; responsive to emergent events
F3	Supports assessment of operational effectiveness
F4	Supports assessment of campaign effectiveness; tracks actual trajectory of OAI against TCO projections
F5	Is able to track past and planned OAI, balance between retrospective and prospective view

similarly participate in the events.[2] Also note that the attention of senior leadership was necessary to move to the reformed JEP: At the outset of the reforms, the J3 requested a review of the command's effects processes and pushed for revision based on that review.

Note that this characteristic—the attention from senior leaders—is distinct from other features and characteristics included in the list. While all the other characteristics and features are something done or included in the process and stem from the functioning of the JEP, "receives senior leader attention" is external to the process that senior leaders carry out. (The design and activities of the process do not determine whether senior leaders attend and participate.) The process can be more or less engaging for senior leaders or take greater or less advantage of their participation and input, but the sense of importance of the process can only come from leaders themselves. This distinction does not make the attention of senior leadership any less important to the success of the JEP.

C2. Centralizes Planning

Centralized planning with decentralized execution is a core tenet of U.S. military doctrine.[3] However, planning can be centralized at one layer in nested HQ and still not be centralized at a higher echelon. The integration of information or OIE is much more likely to be effective when planning is elevated and centralized *at the GCC level* and not left primarily to the SCCs. While the SCCs have a role in planning and execution, GCC battle rhythm events should not just be a place where components' plans are reported and deconflicted but rather where they are developed and integrated, with the final details perhaps worked by service-specific staffs at the components.

C3. Elevates Information in Command and Component Processes

For information and OIE to be effectively integrated in a command's activities, information must be one of the capabilities and effects that is considered and prioritized relative to a command's habitual physical activities. This applies at the GCC level and in the components and requires that information capabilities and effects be elevated and prioritized rather than habitually ignored or underemphasized. This should extend to broadening staff's awareness of the full range of possible capabilities and OAI, including both kinetic and nonkinetic capabilities and efforts. Elevation is needed because the information has gained prominence recently, and it will take time before the concept is fully understood, embraced, and routinized across the force. Until that time, GCCs must begin to realize the vision of integrated physical and informational power by elevating and emphasizing information considerations

[2] The importance of communicating priorities and leadership by example might seem like conventional wisdom, but, as will be discussed, the physical presence of leaders and their active engagement with the JEP processes were repeatedly cited by interviewees as key reasons for their continued improvement. For a discussion of leadership by example (including modeling behaviors in required events), see Michael Schrage, "Like It or Not, You Are Always Leading by Example," *Harvard Business Review*, October 5, 2016.

[3] See also, Clint Hinote, *Centralized Control and Decentralized Execution: A Catchphrase in Crisis?* Air Force Research Institute, March 2009.

in B2C2WG processes that were originally designed primarily for planning and employing physical power.

C4. Involves the Right People in the Right Numbers

One way a command demonstrates commitment to an idea is by allocating resources, and human capital is a critical resource. An effective JEP should involve the right personnel and in the right quantities. Who and how many are right is subjective and holistic and depends on context. Collectively, the *right people* will include the needed expertise; represent the needed components, elements, and directorates; have sufficient representative authority; and include individuals with seniority appropriate to the level of the B2C2WG. This connects to the feature of involving senior leaders: When senior leaders periodically attend boards and bureaus, the staff leads who provide input to those events tend to prioritize the events and send the right personnel. The *right numbers* must balance between having too many personnel in a WG to get any work done and having too few active participants with sufficient expertise to complete the work that is needed. Also, the number of active participants versus the number of occasional, silent, or back-row attendees also affects this balance. Although there is no clear, prescriptive advice on exactly who and how many should be involved, participants in individual battle rhythm events will always know whether the right persons representing certain sections or components—or having needed expertise—are missing.

Category A: Aligns Activity and Strategy

This thematic category includes five features: (A1) promotes a campaigning mindset, (A2) aligns or links OAI with command objectives, (A3) generates unified, synchronized, justified, and well-timed force planning requirements, (A4) demands justification of habitual and routinized events and layers additional OAI to support compulsory events in ways that better tie to theater campaign and IMOs, and (A5) supports or bridges a medium-scale planning time horizon (between CUOPS and FUOPS).

A1. Promotes a Campaigning Mindset

The first characteristic of a JEP that aligns a campaign's activities with its strategy in the current strategic context is that it promotes a campaigning mindset. *Campaigning* refers to sequencing activities in time and space to achieve effects over time. Strategic guidance directs DoD (including the GCCs) to adopt a "campaigning mindset," aligning all activities toward strategic goals.[4] While DoD has recently started applying this idea to contemporary strategic competition, it is not new. Prussian military theorist Carl von Clausewitz described military strategy as linking battles to bring about the objective of the war. DoD is simply extending this long-understood concept to its current environment. On most days, GCC staffs are not

[4] See, for example, Joint Doctrine Note 1-19, *Competition Continuum*, Joint Chiefs of Staff, June 3, 2019, p. 6.

trying to sequence physical engagements with an enemy army to bring about its destruction but working to define effects, plan OAI against those effects, and conduct them in a way that helps achieve incremental progress toward U.S. strategic objectives. The idea of campaigning is relevant at both the competition and conflict ends of the operational continuum and applies across all domains.

A campaigning mindset requires long-term objectives and plans to make incremental contributions to progress toward those objectives from short-, medium-, and long-term OAI. Provocations by competitors and adversaries must been seen in the context of a campaign, not just as crises or episodes, and actions in response or retaliation must be considered for their campaign implications. In addition to the short-term effects on decisions and behavior, the longer-term effects on enduring perceptions and cumulative narratives must be considered in information effects. A campaigning mindset requires structures and processes that connect the short, medium, and long term while sustaining a persistent long-term focus.

A2. Aligns Operations, Activities, and Investments with Strategic Objectives

One of the main functions of a GCC's battle rhythm is to help the command plan and execute activities grounded in its strategic objectives. However, the span of a GCC's control (overseeing the component commands as well as a myriad of joint HQ and TFs) creates a scale and scope of activity that is difficult to manage and align. Because the bulk of operational planning rightfully happens at the component commands, a GCC could struggle to match top-down efforts with bottom-up efforts in a way the ensures that all the OAI happening in the AOR proceed from or are clearly connected to the TCP. These challenges can be exacerbated by the slow burn of the resourcing process in which resourcing and force management decisions made previously (and often prior to the current commander and staff's assumption of duties) can further reduce GCC influence over OAI. Regardless of these impediments, an effective JEP can promote and enforce the explicit alignment and connections between the OAI happening in the AOR and the command's objectives as articulated in its TCP and TCO.

A3. Better Defines Requirements and Demand for Resources

Effective JEPs generate unified, synchronized, justified, and well-timed force planning requirements for all forces, including information forces and capabilities. A large portion of the work of a GCC, as a force employer, is defining and identifying operational requirements necessary to meet strategic objectives and advocating for resources from force providers to meet them. This is not an easy task because it involves planning across lengthy time horizons to align planning, programming, budgeting, and execution processes; force generation cycles; and the timing of external events. Resource requests should be prioritized to match the command's priorities and be unified across components and elements of the command. Resource requests should be thoughtfully and accurately justified, especially when competing with other strategic priorities in other areas of the globe for specific or limited resources or capabilities. Planning for resource requests should align with the deadlines and timelines for input to the larger department-wide resource allocation and planning processes of which they are a part. Resource planners should avoid prioritizing resources simply because they were previously prioritized or constraining requests to resources that are known to be available when other resources are required to achieve desired ends.

A4. Demands Justification of Habits

Large, hierarchical organizations tend to favor efficiency and can function based on inertia and remain resistant to change. Some OAI take place because they have taken place previously, and it is easy to repeat them. Some OAI take place because someone expects that they will (those expectations might come from partners or allies, interagency elements, or civilian staff). Some of these OAI become traditions, and if enough organizations and stakeholders expect them to happen, they might become compulsory (such as an annual exercise that, if canceled, would offend partners expecting to participate). Effective GCC B2C2WG demand justifications for habitual or routinized events, making sure that the reasons that such events were created still apply and still align with the command's objectives. Furthermore, if OAI (or, more likely, an annual event) are both compulsory and not particularly well tied to current TCP or TCO objectives, then an effective JEP would seek to layer additional OAI onto the compulsory activity in ways that better tie effects, especially informational effects, to the TCO and IMOs.

A5. Bridges Planning Across Current and Future Operations

Most military staffs follow a traditional staff structure and separate plans and operations (in a J5 and a J3, for example). The same sort of division is common in the operations directorate in which CUOPS is generally separated from FUOPS within operations (as a J33 and J35). This is the case at most GCCs. However, given the long time horizon of GCC operations, the breadth of the commanders' decision cycle, the time horizons needed for some implementations of information power to have effects, and the need to balance small increments of progress toward IMOs with longer-term, sustained advances, there is a need for the effects process to span the thinking done in the J5 and the J3 and across different operational timelines within the J3.[5] Thus an effective JEP that integrates information and physical power should support a medium-scale planning time horizon and bridge between CUOPS and FUOPS.

Category I: Integrates, Synchronizes, Synergizes, and Layers

This thematic category also includes five features: (I1) is able to generate CONOPS that support multiple objectives and levels and able to coordinate multiple OAI against a single effect or objective; (I2) deconflicts and coordinates authorities and permissions across the service components and other elements; (I3) promotes and enables information sharing across relevant staff sections, components, and capabilities; (I4) integrates with intelligence such

[5] Military officer with expertise in planning, in-person interview with the authors, December 5, 2022; retired senior military officer with expertise in OIE, planning, and operations, telephone interview with the authors, November 7, 2022.

that process is intelligence driven and encourages additional intelligence support; and (I5) accounts for and integrates special activities.

I1. Layers Operations, Activities, and Investments and Effects

A JEP that effectively integrates the information joint function always seeks to layer effects to make activities more effectively contribute to TCP and TCO objectives. This extends to generating CONOPS that support multiple objectives at multiple levels (perhaps supporting multiple IMO, and perhaps achieving operational-level objectives) and includes the ability to coordinate multiple OAI to achieve a single effect, or multiple effects to achieve a single objective, which is integrated multidomain effects. Another way to describe this kind of integration is what LtGen Matthew G. Glavy and Eric X. Schaner label as "21st century combined arms" in which the arms combined now explicitly include information; therefore, combined arms come to be about achieving effects and gaining advantages through the combined effects of maneuver, fires, and information.[6]

I2. Coordinates Permissions and Authorities Across Components and Elements

Within a GCC's AOR and span of control, there will be elements, components, and formations with a wide variety of permissions and authorities. Some permissions and authorities are held at the GCC level (or higher), while many are delegated to the components, JTFs, or operational units. While the bulk of operational planning should be done at the component or JTF level, as noted, this planning should be integrated and deconflicted at the GCC level. There are situations where planners at one echelon might desire a capability or effect that—to the extent of their knowledge—is not available or authorized but might be easily and appropriately delivered by capabilities from another component. Alternatively, a JTF staff that is familiar with using a capability in a certain way might be surprised to discover that the approach is not permitted in a certain country or context. An effective JEP should help manage and deconflict various authorities and permissions to maximize operational effectiveness, prevent surprises, and otherwise deconflict and coordinate authorities and permissions across SCCs and other subordinate elements.

I3. Promotes Information Sharing

As noted, one of the barriers to the effective integration of the information joint function and OIE in broader planning is an unfamiliarity with and lack of awareness of information OAI and related capabilities. Most of the joint force is familiar with a wide variety of military capabilities and aware of their potential contribution to traditional combined arms (fires and maneuver). Many in the joint force are not sufficiently familiar with and aware of information-focused forces and capabilities, or the ways in which military capabilities can be

[6] Matthew G. Glavy and Eric X. Schaner, "21st-Century Combined Arms: Gaining Advantage Through the Combined Effects of Fires, Maneuver, and Information," *Marine Corps Gazette*, September 2022.

leveraged for their inherent informational aspects.[7] This is doubly true for highly technical and compartmented capabilities, such as special activities. Lack of familiarity with the ends, ways, and means of information as a form of power and source of effects can be general, but it can also be specific: Even those who understand the potential contributions of an information joint function might be unaware of important information efforts that are ongoing elsewhere in the command.

Therefore, a set of B2C2WG and processes intended to better integrate the information joint function needs to promote and enable information sharing across relevant staff sections, components, and capabilities to help overcome unfamiliarity with the ends, ways, and means of OIE. No longer can information and the IE be treated as a separate stovepipe that is isolated from the planning and employment of physical power.[8]

I4. Intelligence-Driven Process That Encourages Additional Intelligence Support

While no one in the joint force would dispute the need for operations to be intelligence-driven, in practice, sometimes countervailing factors could cause OAI at a GCC to deviate from that ideal. Where activities are habitual or compulsory or where activities seem like a good idea but related supporting intelligence is lacking, plans can proceed with limited input from intelligence. A lack of prioritization, relevant collections, and analytical skills in the intelligence community can make meeting intelligence requirements for OIE particularly challenging.[9]

Because intelligence collection, production, and analysis are scarce resources, a GCC must be thoughtful in requesting and prioritizing intelligence. In terms of subject-matter expertise and availability to present to and participate in WGs, intelligence support is also a scarce resource: No one person in the J2 (intelligence) directorate is an expert on all topics and regions within a GCC AOR. Therefore, the more specific and narrowly refined a request for intelligence support to a WG is, the more likely the intelligence directorate will be able to meet that request with a single SME. Getting the right analyst in a WG can substantially improve the extent to which related OAI come to be intelligence driven. Furthermore, asking specific and refined questions in WGs in which intelligence is clearly driving planning is likely to engender further intelligence support because additional support can be specific (focused intelligence requests are more manageable) and is rewarded (members of all staff sections are encouraged when their efforts clearly make valued contributions). This has the potential to create a virtuous cycle that improves both operations and intelligence support.

For these reasons, an effective JEP should be integrated with intelligence such that the process is intelligence driven and includes aspects that encourage additional intelligence support.

---

[7] For more on the inherent information aspects of all military activities, Joint Publication 3-04, 2022.

[8] See the discussion under the subheading "What Happens in the IE Does Not Remain in the IE" in Paul et al., 2018.

[9] See Schwille, Atler, et al., 2020.

I5. Accounts for and Integrates Special Activities

Effective JEPs integrate *all* effects from the various OAI across the GCC, including information effects and effects generated by compartmented or special activities. While this seems self-evident, it is nontrivial in practice. Highly classified activities likely need their own WGs, but there also should be cross-participation in those WGs to other battle rhythm events, perhaps with additional personnel read into special programs or "tear line" shareable statements about efforts or their effects at lower classification levels.[10] Special activities should be accounted for and integrated into other aspects of the effects process, including assessment and layering effects to support objectives.

Category F: Provides Feedback and Promotes Iterative Improvement

The final thematic category emphasizes feedback and assessment and includes five features: (F1) is able to provide effective feedback or input to the TCP and TCO process and promote clear IMOs; (F2) supports dynamic or evolving objectives between iterations of TCO cycles and is responsive to emergent events; (F3) supports assessment of operational effectiveness; (F4) supports assessment of campaign effectiveness and able to track the actual trajectory of OAI against TCO projections; and (F5) is able to track past and planned OAI and balance between retrospective and prospective view.

F1. Creates Feedback to Improve Plans and Objectives

To be effective, a JEP should provide feedback and input into the process that develops the command's TCP and TCO, particularly in defining and refining the IMOs that support the overall campaign. A well-run effects process exposes (through contact and friction) whether day-to-day and medium-term efforts easily align or fail to align with the command's IMOs and other objectives. Being able to provide input to the future shape and details of the IMOs allows the staff involved in relevant B2C2WG to affect alignment from both sides: They can adjust the OAI undertaken in pursuit of the objectives, but they can also help adjust the objectives themselves to improve clarity and make it easier to assign tasks that clearly contribute to their completion. Without such feedback, disconnects can persist between activities and objectives, and sufficient clarity and specificity in objectives can be lacking.[11]

---

[10] For more on challenges related to sharing highly classified material within a command, see Martin C. Libicki, Brian A. Jackson, David R. Frelinger, Beth E. Lachman, Cesse Cameron Ip, and Nidhi Kalra, *What Should Be Classified? A Framework with Application to the Global Force Management Data Initiative*, RAND Corporation, MG-989-JS, 2010.

[11] Clear objectives are a cornerstone of assessment and of operations. If what is wanted is not clear, it is almost impossible to get, let alone measure the extent to which it was obtained. Ideally, objectives at all levels should be *SMART*: specific, measurable, achievable, relevant, and time bound. For more on the SMART criteria and the role of objectives in assessment, see Christopher Paul, Jessica Yeats, Colin P. Clarke, Miriam Matthews, and Lauren Skrabala, *Assessing and Evaluating Department of Defense Efforts to Inform, Influence, and Persuade: Handbook for Practitioners*, RAND Corporation, RR-809/2-OSD, 2014.

F2. Supports Evolving Objectives

Objectives do not remain static, so they should be refined as conditions change. Strategic objectives do not change often, but operational and tactical objectives may change rapidly. IMOs, such as those included in the TCP, are revised in sync with the TCP revision cycle over a somewhat lengthy period. However, OAI aligned to these objectives can span the strategic, operational, and tactical levels. Changes in the environment should influence how these are conducted to keep them aligned with the intent of the IMOs. In essence, the GCC functions on multiple overlapping decision cycles.[12] In addition to ensuring that OAI align with strategic objectives and to providing feedback to influence an objective's shape and expression in ongoing planning cycles and documentation, an effective JEP needs to be able to adjust to support evolving objectives before they are formally captured in the IMOs, TCP, and TCO.

F3. Supports Assessments of Operational Effectiveness

Assessment of operational effectiveness (that is, measuring the extent to which OAI produced the intended effects and met operational objectives) is inherently challenging. These challenges only increase when information effects are involved because many forms of information effects are governed more by human dynamics and cognition than by physics; sometimes, they involve causal logic where steps toward the outcome (or the outcome itself) are hard to observe.[13] Effective assessment requires identifying and collecting measures of effectiveness (MOE) and not just measures of performance (MOP).

Regardless of the challenges, assessment of operational effectiveness is an essential pillar to improving operations and to beginning to measure progress toward objectives. An effective JEP should support and promote assessments of operational effectiveness. This might be accomplished through serving as a forcing function and demanding MOE and MOP as part of plans shared by supporting elements. Such an assessment might be supported through collaborative work in WGs to help planners identify possible measures or by working with different elements and assets to support the collection of measures.

F4. Supports Assessment of Progress Toward Theater Campaign Order Projections

Good assessment of OAI is a prerequisite and a building block for good assessment of overall campaigns. While OAI assessment should be owned by the executing elements (though demanded and supported by the GCC battle rhythm), campaign and IMO assessment should be owned by the GCC (supported by data from other elements). An effective JEP should be able to assess the effectiveness of effects in generating progress on campaigns. This should include being able to track the contributions and progress of OAI against projections made as part of the TCO and compare the trajectory of actual progress against expected progress.

---

[12] Military officer with expertise in assessment, in-person interview with the authors, September 16, 2022.

[13] Paul et al., 2014.

Such assessment can contribute to accountability, but it is even more valuable in refining plans and timelines and improving the design and execution of OAI.

F5. Allows for Tracking and Sequencing of Operations, Activities, and Investments in Time and Space

Closely related to assessing progress is the need to track OAI in time and space. This contributes to maintaining situational awareness across the staff regarding ongoing operations and understanding how they relate to one another, the recent past, and future efforts. Part of successful integration involves being sure that intended activities have taken place, and layering effects often require that events unfold in a certain sequence. Note that this involves both *retrospective* (what OAI have taken place) and *prospective* (what OAI are coming) awareness. Ideally, a JEP will find a good balance between reporting out and tracking past accomplishments and sequencing and deconflicting upcoming events.

In Chapter 4, we demonstrate the application of this framework and all 19 of its characteristics or features in a before-and-after assessment of USCENTCOM's battle rhythm events: the pre-JEP B2C2WG (in approximately 2020) and the B2C2WG of the USCENTCOM JEP (approximately 2023).

CHAPTER 4

Assessment and Analysis of U.S. Central Command Battle Rhythm Events

We used the framework described in the preceding chapter to assess USCENTCOM's JEP and to compare it with the command's prior set of battle rhythm events. We conducted a thematic analysis by binning data from interviews, observation, and historical case studies according to the corresponding attributes in the framework. For example, comments by an SME about how the JEP has improved assessment and areas where assessments need to be improved were grouped together. Case study insights about the value of layering effects were synthesized with SME impressions about how the JEP allows planners to synchronize multiple OAI to achieve better outcomes in the IE. By mapping insights against an analytic framework, we were able to identify the strengths of the revised USCENTCOM JEP and highlight areas where there are opportunities for further improvement. Not only is this assessment potentially useful to support continued improvement in USCENTCOM battle rhythm events, but the framework also enables a repeatable assessment process that can be used to evaluate future evolutions of USCENTCOM's battle rhythm events or the B2C2WG of other GCCs with an eye toward improvement.

What Works in U.S. Central Command Battle Rhythm Events?

Overall, USCENTCOM's revised JEP made substantial improvements on the characteristics and features highlighted in the framework when compared with its previous B2C2WG. However, opportunities for further improvement remain.

Table 4.1 provides a summary of the comparison and assessment of the before-and-after (pre-2020 versus 2023) USCENTCOM B2C2WG and battle rhythm events related to information and the JEP. Each table row lists a criterion from the framework described in Chapter 3, an assessment of the 2020 B2C2WG process against that criterion, and an assessment of the 2023 B2C2WG process. In the remainder of the chapter, we discuss each feature or characteristic and the way it is or is not present, followed by a discussion of potential focus areas for improvement.

TABLE 4.1

Before-and-After Comparison of the Boards, Bureaus, Centers, Cells, and Working Groups Processes of the U.S. Central Command Joint Effects Process

Characteristic	Assessment: Pre-2020 B2C2WG	Assessment: 2023 B2C2WG
Centralizes and elevates information power and related effects		
C1. Receives significant senior leader attention	Occasionally	Routinely and importantly
C2. Centralizes and elevates planning rather than leaving planning at the SCCs	Rarely	To a much greater extent
C3. Elevates prominence of information in command processes and in SCC awareness; broadens staff awareness across the range of kinetic and nonkinetic OAI	Low	Much better, but room to improve
C4. Involves the right people, in the right numbers	To some extent	To a much greater extent
Aligns activity and strategy		
A1. Promotes a campaigning mindset	Not really	To a much greater extent, but room to grow
A2. Aligns and links OAI with command objectives	To a limited extent	To a much greater extent
A3. Generates unified, synchronized, justified, and well-timed force planning requirements	No	To a much greater extent, aspires to more
A4. Demands justification of habitual and routinized events; takes compulsory events and layers additional OAI to support them in ways that better tie to TCOs and IMOs	Rarely	To a much greater extent
A5. Supports or bridges a medium-scale planning time horizon (between CUOPS and FUOPS)	Not really	Yes
Integrates, synchronizes, synergizes, and layers informational and physical power		
I1. Is able to generate CONOPS that support multiple objectives/levels; able to coordinate multiple OAI against a single effect/objective	Sometimes	Much more regularly
I2. Deconflicts and coordinates authorities and permissions across SCCs and other elements	Some	To a greater extent
I3. Promotes/enables information sharing across relevant staff sections, components, and capabilities	Some	Much better
I4. Integrated with intelligence such that process is intelligence driven **and** encourages additional intelligence support	Intelligence involved, but too little	Much better, still room for improvement
I5. Accounts for and integrates special activities	To some degree	Much better

Table 4.1—Continued

Characteristic	Assessment	
	Pre-2020 B2C2WG	2023 B2C2WG
Provides feedback and promotes iterative improvement		
F1. Is able to provide effective feedback and input to TCP and TCO process and promote clear IMOs	Rarely	Some, more work needed
F2. Supports dynamic/evolving objectives between iterations of TCO cycle; responsive to emergent events	Limited	Improving and aspires to more
F3. Supports assessment of operational effectiveness	Limited	Improving
F4. Supports assessment of campaign effectiveness; tracks actual trajectory of OAI against TCO projections	No	Improving
F5. Is able to track past and planned OAI, balance between retrospective and prospective view	No	Improving

SOURCE: RAND thematic analysis of data from semistructured interviews, case studies, and observation of battle rhythm events.

NOTE: Each assessment takes the form of a brief commentary and a color-coded stoplight assessment of the presence or absence of the criterion feature or characteristic, with red denoting "predominantly absent," yellow denoting "partially present," and green denoting "substantially present."

Virtues of the Process: Areas Where Desired Attributes Are Substantially Present

Our research indicated that out of the 19 attributes of an effective battle rhythm in our framework, nine of them were substantially present in the current version of the JEP, as opposed to pre-2020 processes in which none of these attributes were substantially present. This research suggests that the JEP has improved USCENTCOM's battle rhythm and highlights those areas where the new processes are especially helpful. Some lessons related to these positive outcomes could potentially be applicable to other contexts (such as other GCC, component, or JTF staffs). The command should consider why it has seen improvements in these areas and continually reassess to determine if continuity or adjustment is needed to sustain this success. The before-and-after assessment of each characteristic begins with the nine features that the revised USCENTCOM JEP has as substantially present. We then turn to the remaining ten characteristics where continued improvement should be desired.

C1. Receives Senior Leader Attention

The JEP has greatly improved the flow of information to senior leaders to support decision-making. Senior leader engagement was critical to improving the JEP and remains an important element of the JEP's ongoing functioning.[1] According to multiple SMEs, senior leader

[1] Interview with a former military officer, video call with the authors, November 7, 2022; military officer with expertise in operations and planning, in-person interview with the authors, September 15, 2022.

buy-in and participation were key factors in the revised JEP's success.[2] The JEP reforms were initiated by the USCENTCOM J3, which, based on experience in Operation Inherent Resolve and ongoing work at the Joint Staff, saw a need for an effects-based approach.[3] The JEP adaptations had the endorsement and support of the USCENTCOM commander at the time, a level of buy-in that appears to have remained across changes of command.[4] Along with providing "top cover" to allow the J3 section to adapt its processes, the previous USCENTCOM commander displayed interest by attending the JEWG, reportedly calling it "the best brief he had received while in command."[5] USCENTCOM and component commander participation was also central to such events as the PRAC, in which they are the primary participants, and the SAOC, in which their engagement allows for a synthesis of special activities with other OAI that had heretofore not occurred at the command.[6] JEP battle rhythm events also received participation and buy-in from key personnel, such as the deputy J3 and chief of staff, who attended prescribed meetings and sometimes sat in on other meetings for situational awareness and support.[7] Prior to 2020, high-level participation and interest in battle rhythm events were much more sporadic.

C2. Centralizes Planning

Under previous processes, planning was often left to service components with USCENTCOM sitting above them as an additional layer.[8] The JEP elevated substantive planning to the GCC level when appropriate and allowed USCENTCOM staff to give guidance to the components while fostering coordination across them. According to several of the SMEs we interviewed, the GCC's ability to plan strategically and the components' ability to execute based on these higher-level plans have improved substantially with the adoption of the JEP.[9] RAND's obser-

---

[2] Interview with a former military officer, video call with the authors, November 7, 2022; military officer with expertise in operations and planning, in-person interview with the authors, September 15, 2022.

[3] Interview with a former military officer, video call with the authors, November 7, 2022.

[4] Interview with a former military officer, video call with the authors, November 7, 2022.

[5] Military officer with expertise in plans, in-person interview with the authors, September 15, 2022.

[6] Military officers with expertise in planning, in-person interview with the authors, December 5, 2022; military officer with expertise in assessment, in-person interview with the authors, September 16, 2022; retired senior military officer with expertise in OIE, planning, and operations, telephone interview with the authors, November 7, 2022.

[7] Military officer with expertise in operations and planning, in-person interview with the authors, September 15, 2022; retired senior military officer with expertise in OIE, planning, and operations, telephone interview with the authors, October 5, 2022.

[8] Military officers with expertise in planning and operations, in-person interview with the authors, September 14, 2022; retired senior military officer with expertise in OIE, planning, and operations, telephone interview with the authors, October 5, 2022.

[9] Retired senior military officer with expertise in OIE, planning, and operations, telephone interview with the authors, November 7, 2022; military officers with expertise in planning and operations, in-person inter-

vations from the JEB and JEWG, in which components and subordinate TFs participated, largely support this assertion.[10]

C4. Involves the Right People, in the Right Numbers

One of the virtues of the JEP is that it provides the right touch points, and each event is appropriately scoped to the right number of participants.[11] For example, the JEWG is the largest forum because it is intended to build situational awareness and catalyze follow-on collaboration. The IEWG is more focused on the refinement of CONOPS from the components. The SAWG is different in that it involves only a small number of people. While some participants are concerned that relevant players are sometimes excluded, many of the key planners involved also attend other battle rhythm events to achieve the right trade-off between operational security and situational awareness.[12]

Prior to the reform of the JEP, it was difficult to get the right people together for collaboration. Action officers would often spend a long time trying to locate the right point of contact for an issue, and staff sections were at a great risk of stovepiping: remaining aware of their own key issues without understanding relevant information outside their immediate organization.[13]

A2. Aligns Operations, Activities, and Investments with Strategic Objectives

Prior to the revised JEP's implementation, the extent to which OAI undertaken within the USCENTCOM AOR aligned to strategic objectives was unclear. This was complicated by the shift in strategic objectives within national policy and the drawdowns in Iraq and Afghanistan.[14] Multiple staff officers explained to the research team that some OAI were occurring only because they were planned long in advance or because similar things were done in the

view with the authors, September 14, 2022.

[10] USCENTCOM, *CENTCOM Joint Effects Working Group*, September 1, 2022b, Not available to the general public; USCENTCOM, *CENTCOM Joint Effects Working Group*, September 15, 2022c, Not available to the general public; USCENTCOM, *CENTCOM Joint Effects Board Information Flow*, March 15, 2022a, Not available to the general public.

[11] Retired senior military officer with expertise in OIE, planning, and operations, telephone interview with the authors, November 7, 2022.

[12] Military officer with expertise in operations and planning, in-person interview with the authors, September 15, 2022.

[13] Retired senior military officer with expertise in OIE, planning, and operations, telephone interview with the authors, November 7, 2022; military officer and civilian with expertise in PSYOP and planning, in-person interview with the authors, December 6, 2022.

[14] SME, interview with the authors, September 15, 2022.

past.[15] Other OAI were undertaken because there was an opportunity to do so (capabilities were available) or because they seemed like good ideas.[16] The adoption of the JEP greatly improved the connections between OAI and objectives. For example, the PRAC allows for cross-component visibility and a methodical ranking of all OAI. This helps ensure that each element of OAI proceeds from and connects to an IMO in the TCP.[17] During medium-term planning and execution, the JEP helps USCENTCOM staff ensure that what they are doing is grounded in the original intent. For example, the IEWG does this by producing "better CONOPS," meaning that they are developed with informational effects in mind.[18] The JEWG forces a conversation about recent, ongoing, and upcoming OAI, which encourages the staff to clarify and refine their desired effects.[19] Similarly, the JEB brings priority effects before senior leaders for visibility and feedback in the form of guidance and direction.[20]

A5. Bridges Planning Across Current and Future Operations

In most military staffs, plans, and operations are strictly bifurcated (such as in a J5 and a J3). The same is true for operations, where CUOPS is generally separated from FUOPS within the Operations Directorate (in a J33 and a J35). This is true at USCENTCOM as it is at most GCCs. Given the long time horizon of GCC operations and the breadth of the commanders' decision cycle, the JEP acknowledges the need for the effects process to span the thinking done in the J3 and the J5.[21] Such events as the PRAC and campaign assessments reside in plans and the IEWG, JEWG, and JEB reside in operations.[22] The J39 acts as the synchronizer for the entire JEP, bridging the gap between CUOPS and FUOPS. This gives the staff a medium-term look at the alignment between activities that are recently completed, upcoming, and underway to understand how they align to higher-level objectives and to one another. This perspective was predominantly absent in USCENTCOM B2C2WG prior to the adoption of the revised JEP. In

[15] Military officers with expertise in planning, in-person interview with the authors, December 5, 2022.

[16] Military officers with expertise in planning, in-person interview with the authors, December 5, 2022.

[17] Senior military officers with expertise in OIE and planning, in-person interview with the authors, December 5, 2022; USCENTCOM, 2022c.

[18] Military officers with expertise in planning and operations, in-person interview with the authors, September 14, 2022; military officer and civilian with expertise in PSYOP and planning, in-person interview with the authors, December 6, 2022.

[19] Senior military officers with expertise in OIE and planning, in-person interview with the authors, December 5, 2022.

[20] USCENTCOM, March 15, 2022a.

[21] Military officer with expertise in planning, in-person interview with the authors, December 5, 2022; retired senior military officer with expertise in OIE, planning, and operations, telephone interview with the authors, November 7, 2022.

[22] Military officer with expertise in assessment, in-person interview with the authors, September 16, 2022; military officers with expertise in planning and operations, in-person interview with the authors, September 14, 2022.

the view of interviewed SMEs, the JEP offers a look at "where we are at in this plan."[23] This can help the staff bring activities into line with the TCP while providing feedback to inform future planning and even the TCO itself. Staff SMEs suggested that as much as the JEP has helped to establish this feedback loop, some work remains to refine the process and truly get it right.[24] Formats remain in flux, and the staff is still developing practices to get the B2C2WG formats to truly shape both CUOPS and FUOPS.

I1. Layers Operations, Activities, and Investments and Effects

Prior to the JEWG's implementation, planners had little opportunity to layer OAI because of the lack of mid-range visibility during planning and execution.[25] Now, the JEWG provides the command with a forum in which the staff and representatives from the components and subordinate joint HQs can view all recent, ongoing, and upcoming OAI. The conversation in the JEWG centers about what is happening and why has allowed for follow-on refinement of CONOPS or the addition of new activities that take advantage of effects created by other events.[26] This is especially important when integrating informational capabilities. Every OAI will have informational effects, and understanding what this effect is intended to be (or what it inadvertently could be) allows the command to tailor other activities around it or apply other capabilities to amplify the desired effect while mitigating others.

I2. Coordinates Permissions and Authorities Across Components and Elements

The JEP provides a mechanism for components and other subordinate HQ to obtain permissions (for which authority is held at the USCENTCOM level) or apply GCC-retained capabilities. The bulk of operational planning should be done at the component or JTF level, but sometimes, due to risk or resources, a decision is needed from a senior leader. The IEWG provides a venue for injecting new CONOPS and the JEWG for socializing and refining them.[27] The JEB allows priority effects to be elevated to the general officer or flag officer level, so

[23] Retired senior military officer with expertise in OIE, planning, and operations, telephone interview with the authors, November 7, 2022.

[24] Military officer with expertise in assessment, in-person interview with the authors, September 16, 2022.

[25] Military officer and civilian with expertise in PSYOP and planning, in-person interview with the authors, December 6, 2022; retired senior military officer with expertise in OIE, planning, and operations, telephone interview with the authors, November 7, 2022; USCENTCOM, 2022b; USCENTCOM, 2022c.

[26] Retired senior military officer with expertise in OIE, planning, and operations, telephone interview with the authors, November 7, 2022.

[27] Military officers with expertise in planning, and operations, in-person interview with the authors, September 14, 2022.

they can be staffed appropriately.[28] Of course, previous processes allowed subordinate commanders to request approval or additional resources from the GCC. However, it appears that the JEP provides more-effective avenues to raise issues in a timely manner and on a regular periodicity. This further enables the centralized planning and decentralized execution that is desired for an effective battle rhythm.

13. Promotes Information Sharing

The current JEP promotes information sharing across the staff and among the components to a much greater extent than previous processes. One of the most important benefits of the JEB described by USCENTCOM planners was that it brings people together throughout the staff. This allows action officers responsible for one process to understand how it relates to another, or a component planning an exercise or KLE to account for its relationship to ongoing operations by another component. For example, the JEWG has representatives from across the GCC staff and dial-ins from each component and key subordinate HQs. Everyone briefs what they are doing and why, which allows for discussion and follow-on coordination.[29]

15. Accounts for and Integrates Special Activities

The JEP helps USCENTCOM integrate *special activities*, which are focused capabilities that require centralized authorities and permissions at the highest levels and must be protected with extra layers of operational security. Previous processes also had avenues for planning these, but previous versions of what is now the SAWG were underutilized and not well tied into the broader effects process.[30] Special activities were not tailored as well as they could have been to align with other OAI, and those responsible for integrating OAI at the senior level were limited in their ability to "see the whole picture."[31]

The JEP revitalized the SAWG, and it sits parallel to the IEWG and performs the same function for those sensitive OAI that cannot be briefed to a wider audience.[32] The SAOB sits above the SAWG and meets on a biweekly cadence, whereas the SAWG meets weekly. The SAOB gives more-senior leaders visibility on the activities and an opportunity to refine

[28] Military officer with expertise in planning and operations, in-person interview with the authors, December 5, 2022.

[29] USCENTCOM, 2022b; USCENTCOM, 2022c.

[30] Retired senior military officer with expertise in OIE, planning, and operations, telephone interview with the authors, October 5, 2022.

[31] Military officer with expertise in planning and operations, in-person interview with the authors, December 5, 2022; retired senior military officer with expertise in OIE, planning, and operations, telephone interview with the authors, November 7, 2022; military officer with expertise in operations and planning, in-person interview with the authors, September 15, 2022.

[32] Military officer with expertise in operations and planning, in-person interview with the authors, September 15, 2022.

OAI.[33] Importantly, many of the same action officers that work on the SAWG and the leaders involved in the SAOB are also involved in the IEWG and the JEWG and thus understand the relationship between special activities and other OAI.[34] The biannual SAOC allows the component commanders and the USCENTCOM commander to see all sensitive activities, and make key decisions on programs in the AOR.[35] End-to-end, this process mirrors the rest of the JEP with key points of contact to ensure alignment. The previous processes did not have these mechanisms, and the SMEs we interviewed all suggested that the current JEP more effectively integrates special activities.[36]

Continued Improvement: Areas Where Positive Attributes Are Partially Present

Using our framework for assessing a JEP, we found that the revised JEP left USCENTCOM with opportunities for further improvement on ten of them, with the positive attributes partially present. Prior to the JEP, these desired characteristics were mostly absent in seven of the key areas. This indicates that the adoption of the JEP has significantly improved some of USCENTCOM's persistent problem areas, even if additional improvement is possible and desirable.

C3. Elevates Information in Command and Component Processes

Prior to the current JEP, information did not factor prominently in USCENTCOM planning. The typical JEB at most operational HQ focuses on kinetic targeting, and USCENTCOM was no exception.[37] According to SME input, a typical targeting cycle might have seemed sufficient for USCENTCOM because of 20 years of ongoing war in the theater. The change with the revised JEP was stark. As one staff officer stated, "most effects boards are

[33] Retired senior military officer with expertise in OIE, planning, and operations, telephone interview with the authors, October 5, 2022; military officer with expertise in planning and operations, in-person interview with the authors, December 5, 2022; military officer with expertise in operations and planning, in-person interview with the authors, September 15, 2022.

[34] Retired senior military officer with expertise in OIE, planning, and operations, telephone interview with the authors, October 5, 2022.

[35] Retired senior military officer with expertise in OIE, planning, and operations, telephone interview with the authors, November 7, 2022.

[36] Retired senior military officer with expertise in OIE, planning, and operations, telephone interview with the authors, October 5, 2022; military officer with expertise in planning and operations, in-person interview with the authors, December 5, 2022; military officer with expertise in operations and planning, in-person interview with the authors, September 15, 2022.

[37] Military officer with expertise in planning and operations, in-person interview with the authors, December 5, 2022; retired senior military officer with expertise in OIE, planning, and operations, telephone interview with the authors, November 7, 2022.

about 'warheads on foreheads,' but here, we recognized that if you do effects solely from a fires perspective, you miss something."[38] When the J3 originally implemented the JEWG, they housed it within the command's fires cell as part of the traditional targeting cycle. This proved an awkward fit and did not place the emphasis on informational effects and the IE that the command intended.[39] To get the desired emphasis, responsibility of the JEWG was moved into the J39 with an explicit information focus. Expertise from such fields as field artillery are still present, but these perspectives are applied within an information-centric process.[40]

As the JEP evolved, the staff realized that the JEWG did not fully replace the functions of the JTWG. In reality, the traditional targeting cycle (both kinetic and nonkinetic) overlaps with and is reinforced by the kind of mid-range, holistic thinking done in the JEWG. At the culmination of the revisions to the JEP documented in this report, the JTWG had been re-established, with many of the same participants as the JEWG to ensure alignment as common issues are discussed.[41] The staff also realized that the effects process could not be divorced from other forms of strategic communication. The military has an understandable desire to strictly separate such activities as public affairs, where the goal is to inform with facts (and the domestic population is a key audience), from activities like MILDEC, which aims exclusively at enemies and adversaries. There are legal, ethical, and operational reasons for this. However, overt communications are a vital aspect of OIE. USCENTCOM Communications Integration Public Affairs now sends public affairs personnel to the JEWG so that they can understand how their efforts align with the broader effects process and inform the staff's thinking regarding the IE.[42]

Despite great progress, the JEP has not yet elevated information in component processes to the extent desired. While it has "helped get the component staffs talking about information," a truly effects-based approach that takes full account of the IE requires a mindset shift.[43] Although action officers now regularly include desired effects in their OAI briefings, it is unclear whether planning truly begins with desired cognitive effects and treats informational effects as coequal with physical ones.[44]

---

[38] Senior military officers with expertise in OIE and planning, in-person interview with the authors, December 5, 2022.

[39] Retired senior military officer with expertise in OIE, planning, and operations, telephone interview with the authors, November 7, 2022.

[40] SMEs with planning and fires expertise, interview with the authors, September 15, 2022.

[41] Senior military officer with expertise in fires and planning, in-person interview with the authors, December 5, 2022.

[42] Strategic communications professionals, interview with the authors, September 16, 2022; military officers with expertise in planning, in-person interview with the authors, December 6, 2022.

[43] Military officers with expertise in planning and operations, in-person interview with the authors, September 14, 2022.

[44] USCENTCOM, 2022b; USCENTCOM, 2022c.

A1. Promotes a Campaigning Mindset

The JEP helps USCENTCOM foster a campaigning mindset. The current process is far better suited to defining effects and planning against them over time than previous processes, but there is still room to grow.[45] This is the realm of operational art, which, as the name suggests, is not an exact science. There is no one answer to how a GCC should structure itself to carry out a campaign, and USCENTCOM is still refining its B2C2WG to better embrace campaigning. Part of the challenge in this regard is USCENTCOM's recent history as a warfighting theater. With widespread kinetic operations in Iraq and Afghanistan, planners were often focused on near-term effects in the physical domain, such as kinetic strikes on terror networks.[46] The new strategic reality requires USCENTCOM to think longer-term about tensions with Iran, the persistent threat of violent extremist organizations, and strategic competition with China and Russia. These tasks are less kinetic and more cognitive.[47] The AOR also sees fewer deployments, placing an even greater premium on the ability to achieve effects with limited resources.[48]

The JEP was born from the realization that all GCC OAI were really about influencing someone's will to change their behavior. In the words of one SME, the point of the JEP was to "think beyond creating a smoking hole."[49] This could be deterring a country from military aggression, decreasing a terrorist group's ability to plan attacks, or assuring an ally that the U.S. military is a reliable partner. The JEP helps bridge the gap between the campaign plan and the actual OAI that USCENTCOM carries out, which is operational art.[50] The staff is now thinking beyond doing an activity to why that activity is being done and how it relates to other actions the command is taking. There are still barriers, such as the tendency of near-term operations to dominate attention and balancing the time it takes to refine the campaign plan while planning and resourcing effects-based OAI.[51] Interviews with SMEs and

---

[45] Retired senior military officer with expertise in OIE, planning, and operations, telephone interview with the authors, November 7, 2022.

[46] Senior military officers with expertise in OIE and planning, in-person interview with the authors, December 5, 2022.

[47] USCENTCOM civilian with expertise in OIE and planning, in-person interview with authors, September 15, 2022.

[48] USCENTCOM civilian with expertise in OIE and planning, in-person interview with authors, September 15, 2022.

[49] Senior military officers with expertise in OIE and planning, in-person interview with the authors, December 5, 2022.

[50] Senior military officers with expertise in OIE and planning, in-person interview with the authors, December 5, 2022.

[51] Retired senior military officer with expertise in OIE, planning, and operations, telephone interview with the authors, November 7, 2022; military officer with expertise in planning and operations, in-person interview with the authors, December 5, 2022.

RAND's case studies revealed that USCENTCOM faces persistent challenges in adapting to campaigning but made great strides between 2020 and 2023.[52]

A3. Defines Requirements and Demand for Resources

As force employers, a large portion of a GCC's work is defining operational requirements and advocating for resources from force providers to meet them. Previously, USCENTCOM faced two specific related challenges. First, with two or more active conflicts in the AOR for much of the previous two decades, the command was flush with resources and often offered or assigned capabilities in ways that one SME characterized as "just getting stuff and then finding ways to use it," which led toward thinking about what can be accomplished with the capabilities available rather than designing intended effects and requesting resources to match.[53] Second and related, given the high operating tempo faced within in the command (and by the staff), some resource requests became habitual, duplicating historical requests.[54]

The adoption of the revised JEP has fostered a mindset shift away from this counter-productive tendency toward one in which planners define requirements based on the effects they want to achieve and find creative ways to resource them with what is available.

The PRAC is key in this regard because it is the most forward-looking battle-rhythm event. It accounts for PPB&E as well as Global Force Management while providing component commanders a means to advocate for their priority OAI.[55] OAI are rank-ordered for resourcing with the broader context in mind.[56] This feeds the rest of the JEP, because USCENTCOM's priorities are grounded in effects, and the resources coming into theater are based on better-defined requirements. The PRAC allows USCENTCOM to look at the PPB&E and global force management (GFM) cycles to identify opportunities and risks, and then generate requests for forces that will help the command better achieve its objectives.[57] The PPB&E and GFM cycles are complex and outside the control of a GCC. Because the PRAC is biannual and looks six months ahead, it takes time to actually use the process to its fullest extent and identify meaningful windows of opportunity.[58] USCENTCOM planners

[52] Senior military officers with expertise in OIE and planning, in-person interview with the authors, December 5, 2022.

[53] Military officers with expertise in planning, in-person interview with the authors, December 5, 2022.

[54] Military officers with expertise in planning, in-person interview with the authors, December 5, 2022.

[55] Military officers with expertise in planning, in-person interview with the authors, December 5, 2022.

[56] Military officer with expertise in assessment, in-person interview with the authors, September 16, 2022; military officers with expertise in planning, in-person interview with the authors, December 5, 2022.

[57] Military officer with expertise in assessment, in-person interview with the authors, September 16, 2022.

[58] Military officer with expertise in assessment, in-person interview with the authors, September 16, 2022; military officers with expertise in planning, in-person interview with the authors, December 5, 2022.

acknowledged that there is some way to go before requirements fully drive resourcing rather than the other way around.[59]

A4. Demands Justification of Habits

The JEP helps USCENTCOM to better define requirements and to align OAI with strategic objectives. When this is not happening, the staff can end up caught by inertia and simply repeating what has been done before. In fact, one SME suggested that, prior to the JEP, USCENTCOM was to some extent "asleep at the wheel" in certain areas of the operations it was planning and conducting based on past precedent and available resources.[60] Organizations have cultures, and the good and bad behaviors they practice can be difficult to change.[61] The JEP has not entirely eliminated the tendency toward inertia that makes creative planning difficult at a GCC, nor can it be expected to. However, multiple SMEs asserted that the JEP has improved the situation by checking complacency and encouraging initiative. It does this by forcing conversations in such venues as the JEWG.[62] Planners now have greater visibility on what is happening across the command and a venue where they can question what they are doing and why. This allows CONOPS to be refined prior to execution and the staff to identify opportunities that otherwise may have been missed. For example, someone might notice that an exercise or a deployment occurs at the same time as another OAI and then develop a better CONOPS that takes advantage of the connection between the two. Also, an OAI might be planned without due consideration of its impacts in the IE. SMEs could take this issue for action and develop a CONOPS for informational activities that amplify the desired effects, mitigate risks, or both. In general, the JEP provides more opportunities for the staff to ask "why?" and adjust course during planning.[63]

I4. Encourages Additional Intelligence Support Through Intelligence-Driven Process

Ideally, the JEP would be a truly intelligence-driven process, meaning that OAI are planned based on the command's understanding of the environment and assessed based on observable changes in behavior. Our observation of battle rhythm events and discussions with

[59] Military officer with expertise in assessment, in-person interview with the authors, September 16, 2022; military officers with expertise in planning, in-person interview with the authors, December 5, 2022.

[60] Military officers with expertise in planning, in-person interview with the authors, December 5, 2022.

[61] Bill Gormley, "James Q. Wilson, *Bureaucracy: What Government Agencies Do and Why They Do It*," in Martin Lodge, Edward C. Page, and Steven J. Balla, eds., The Oxford Handbook of Classics in Public Policy and Administration, Oxford, 2015.

[62] Retired senior military officer with expertise in OIE, planning, and operations, telephone interview with the authors, November 7, 2022.

[63] Senior military officers with expertise in OIE and planning, in-person interview with the authors, December 5, 2022.

SMEs have led us to believe that the JEP has improved, and continues to improve, the use of intelligence in planning. There is, however, a long way to go to fully integrate intelligence into planning and assessment.[64]

The JEWG, for example, includes a briefer from the J2 that sets the stage for subsequent discussion and speaks holistically to an LOE or region based on the battle rhythm. This detailed overview brief helps contextualize the discussion about ongoing OAI, their relationship to one another, and desired effects.[65] It is difficult, however, to integrate intelligence in a way that actively informs the discussion, which would require briefers with the exact right expertise and likely follow-on work beyond the JEWG.[66] The J39 also benefits from human terrain analysts who help inform effects-based planning through a more sophisticated view of regional trends.[67] Ties with the broader USCENTCOM J2 are still maturing, with the aspiration to reach a point at which CONOPS are truly informed by analytic judgments and SMEs actively engage in discussion about ongoing OAI.[68]

Assessment would benefit from more-focused intelligence support. While the JEP is maturing the use of well-defined MOE and MOP to measure operations and campaign progress against, these measurements could be better informed by intelligence. As effects are related to behavior change, intelligence is one of the primary means of understanding what kinds of behavior changes are or are not taking place among target audiences. The J39 needs a greater ability to influence collection requirements. There may be opportunity here because USCENTCOM possesses a robust intelligence enterprise that is reorienting from an era of persistent conflict and kinetic operations to the current environment of competition in which understanding the behavior of target audiences is more important than planning strikes or assessing battle damage.[69]

F1. Creates Feedback to Improve Plans and Objectives

The JEP improved USCENTCOM's ability to align OAI with higher-level objectives and assess both operational effectiveness and progress toward those objectives. However, effects-

---

[64] USCENTCOM, 2022c; senior military officers with expertise in OIE and planning, in-person interview with the authors, December 5, 2022; retired senior military officer with expertise in OIE, planning, and operations, telephone interview with the authors, November 7, 2022; civilians with expertise in social media activities and planning, in-person interview with the authors, December 6, 2022; civilians with expertise in intelligence, in-person interview with authors, September 16, 2022.

[65] USCENTCOM, 2022c.

[66] Senior military officers with expertise in OIE and planning, in-person interview with the authors, December 5, 2022.

[67] Civilians with expertise in intelligence, in-person interview with authors, September 16, 2022.

[68] Civilians with expertise in intelligence, in-person interview with authors, September 16, 2022.

[69] Civilian with expertise in assessments, in-person interview with authors, December 6, 2022; senior military officers with expertise in OIE and planning, in-person interview with the authors, December 5, 2022; civilians with expertise in intelligence, in-person interview with authors, September 16, 2022.

based planning grounded in the TCP and assessment (both operational and of the campaign) remain two of USCENTCOM's most vexing challenges. Part of the challenge is that military objectives are not static. The TCP changes on a relatively lengthy time horizon, and while the JEP has improved mid-range planning, it has only marginally improved feedback into plans and objectives. Ideally, operations would inform the refinement of guidance, such as IMOs, which would improve operational effectiveness while better enabling assessment.

The PRAC (where USCENTCOM ranks and prioritizes resourcing for OAI) and the IEWG (where CONOPS for priority effects are injected into the planning process on a 270-day time horizon) have improved alignment with higher-level objectives but have not fully established a feedback loop with them.[70] According to some SMEs, the campaign plan itself remains the biggest problem.[71] Both effects-based planning and assessment depend on clear objectives.[72] It is difficult to plan and assess operations because IMOs are often vague and abstract in the campaign plan.[73] While the JEP has helped bridge CUOPS and FUOPS and better aligned OAI with higher-level guidance, lessons from planning and execution are not currently informing the refinement of higher-level plans and objectives.

F2. Supports Evolving Objectives

While IMOs are revised on a fixed cycle, the JEP promotes opportunities to respond to evolving objectives. The JEP allows CONOPS to enter the decision cycle (via the IEWG) on a medium-term planning horizon, the JEWG provides visibility (on about a 90-day range), and the JEB facilitates senior leader direction and guidance back into the cycle to gain critical approvals or refine OAI.[74] In this way, the staff and components are getting the opportunity to continually evaluate their efforts and possibly make adjustments. As conditions change, the desired effects may be modified at the margins, or new opportunities for messaging on an emergent issue can be seized.

F3. Supports Assessment of Operational Effectiveness

Operational assessment is inherently challenging. It becomes even more so when adopting an effects-based approach. The SMEs we interviewed emphasized that USCENTCOM realizes that the use of military power is about changing behavior, and this means that the most con-

---

[70] Military officers with expertise in planning and operations, in-person interview with the authors, September 14, 2022; military officer with expertise in assessment, in-person interview with the authors, September 16, 2022.

[71] Military officer with expertise in planning and operations, in-person interview with the authors, December 5, 2022.

[72] Military officer with expertise in assessment, in-person interview with the authors, September 16, 2022.

[73] Military officer with expertise in planning and operations, in-person interview with the authors, December 5, 2022.

[74] USCENTCOM, 2022.

sequential effects are cognitive.[75] This is even more apparent now that USCENTCOM is conducting fewer kinetic operations while trying to make the most out of limited resources.[76] J39 has established an assessment process and evaluates specific OAI based on MOE and MOP. The JEP facilitates this, and there is now an SME in charge of assessments located within the branch running the process.[77] It also promotes the defining MOP and MOE throughout the process, from the IEWG when CONOPS are introduced to the JEWG and JEB where everything is presented in terms of MOE.[78] Such tools as the C2IE software application help planners catalog MOP and MOE for OAI and add assessment data after the fact.[79]

Operational assessment within the JEP is still in its early stages. While planners are becoming habituated to briefing in terms of MOP and MOE, it does not guarantee that all planning truly begins with the desired cognitive effect.[80] Such tools as C2IE are helpful, but are only as useful as the data entered into them.[81] MOE are particularly hard to define, and if those doing the planning or data entry have not considered them carefully, it limits the utility of any kind of assessment.[82] Planners must also adopt the habit of entering complete and detailed data after an event and continually referencing the application to inform ongoing efforts. We were able to view J39 assessment products that demonstrated effects-based thinking. SMEs were unanimous in their view that assessment has gradually improved in the JEP process but remains one of the most challenging areas.[83]

[75] Senior military officers with expertise in OIE and planning, in-person interview with the authors, December 5, 2022; USCENTCOM civilian with expertise in OIE and planning, in-person interview with authors, September 15, 2022.

[76] USCENTCOM civilian with expertise in OIE and planning, in-person interview with authors, September 15, 2022.

[77] USCENTCOM SME, in-person interview with the authors, September 16, 2022.

[78] USCENTCOM, 2022b; USCENTCOM, 2022c; USCENTCOM, 2022a.

[79] Military officers with expertise in planning, in-person interview with the authors, December 5, 2022.

[80] USCENTCOM, 2022a.

[81] Military officer and civilian with expertise in planning and analysis, in-person interview with the authors, December 6, 2022.

[82] Civilians with expertise in social media activities and planning, in-person interview with the authors, December 6, 2022; military officer with expertise in assessment, in-person interview with the authors, September 16, 2022.

[83] Military officer with expertise in assessment, in-person interview with the authors, September 16, 2022; civilian with expertise in assessment, in-person interview with the authors, September 16, 2022; senior military officers with expertise in OIE and planning, in-person interview with the authors, December 5, 2022.

F4. Supports Assessment of Progress Toward Theater Campaign Order Projections

Previous GCC processes provided little means of assessing operations against the TCP and determining whether USCENTCOM made progress toward its objectives.[84] The JEP has helped bridge the gap between campaign assessment (the biannual analysis of the TCP) and operational assessment (the grading of individual OAI). At the conclusion of our data collection, however, the staff was just getting started in determining the best way to do this going forward.[85] Assessing progress toward the TCP remains one of the biggest challenges for the JEP process.

Responsibility for campaign assessment resides in the J5. The process was moved to the J5 from the fires cell due to the broader scope of the process beyond the immediate impact of kinetic strikes that fires planners are typically accustomed to. This move occurred around the time that the J5 stood up the PRAC process.[86] Now, the IMOs are used as benchmarks for assessing OAI. As of the conclusion of data collection for this project in early 2023, the J5 had completed three campaign assessments under the JEP process, which are a compilation of OAI.[87] In the preceding two years, the section moved to a quarterly process.[88] This process is one of many assessment processes within the command. The J5 starts with the IMOs, which are derived from campaign objectives and associated with lists of desired effects. *Desired effects* are defined as "conditions that result from an act, or conditions that must be achieved."[89] These effects are further broken down to indicators of success, which are the specific and describable elements that tell whether an effect is being achieved. The indicators are ranked from most to least favorable. At the time of this writing, the J5 was using about 200 defined indicators of success for its assessment process.[90]

There are two main challenges with maturing the campaign assessment process. First, the J5 is tasked with asking "how does this relate to the campaign plan?" and the J3 focuses on "how did this [OAI] go?"[91] At the time of this writing, planners were struggling to bridge the gap between the J39 and J5 assessments in a way that could give USCENTCOM an understanding of how it was moving toward its campaign objectives, away from them, or both. The second challenge is that of defining objectives in the first place. Because the indicators that the J5 uses for assessment are derived from the MOE, and the MOE are derived from IMO,

[84] Military officers with expertise in planning, in-person interview with the authors, December 5, 2022.

[85] Military officer with expertise in assessment, in-person interview with the authors, September 16, 2022.

[86] Military officers with expertise in planning, in-person interview with the authors, December 5, 2022.

[87] Military officer with expertise in assessment, in-person interview with the authors, September 16, 2022.

[88] Military officer with expertise in assessment, in-person interview with the authors, September 16, 2022.

[89] Military officer with expertise in assessment, in-person interview with the authors, September 16, 2022.

[90] Military officer with expertise in assessment, in-person interview with the authors, September 16, 2022.

[91] Senior military officer with expertise in OIE, planning, and operations, telephone interview with the authors, October 5, 2022.

an assessment will only be as good as the IMO. If the IMOs are not well grounded in strategic objectives, are poorly defined, or are not SMART, the already difficult task of assessing progress toward the campaign becomes more difficult and potentially loses its value altogether.[92]

F5. Allows for Tracking and Sequencing of Operations, Activities, Investments in Time and Space

Closely related to assessing progress is the need to track OAI in time and space. This means maintaining situational awareness across the staff regarding ongoing operations and understanding how they relate to one another, the recent past, and future efforts. The JEWG has been critical in this regard in that it brings the staff together with representatives from the subordinate HQ to brief their recent OAI. The staff not only briefs OAI but can visually display them in relation to one another and discuss disconnects or potential opportunities.[93]

The staff has also adopted tools to make OAI and their effects easier to track. C2IE software was deployed circa 2019–2020, and its use has gradually increased and expanded across the staff.[94] C2IE helps maintain situational awareness and, as previously discussed, facilitates operational assessment. SMEs reported that this tool supported the JEP by allowing planners to understand what was going on in context without having to "chase an action officer."[95] Together, the JEP process and the associated software tools contribute to the campaigning mindset and the ability to assess that are inherent to an effective battle rhythm.

Building on this before-and-after assessment of the implementation of the revised JEP at USECENTCOM, Chapter 5 offers findings and recommendations.

---

[92] Military officers with expertise in planning, in-person interview with the authors, December 5, 2022; military officer with expertise in assessment, in-person interview with the authors, September 16, 2022.

[93] USCENTCOM, 2022b; USCENTCOM, 2022c.

[94] Military officers with expertise in planning, in-person interview with the authors, December 5, 2022.

[95] Military officers with expertise in planning, in-person interview with the authors, December 5, 2022.

CHAPTER 5

Findings and Recommendations

Findings

Our literature review, semistructured interviews with USCENTCOM staff, historical case studies, and observations of battle rhythm events showed that the JEP has improved USCENTCOM's ability to conduct the information joint function and operate in the IE. Operating in the IE is inherently difficult, and the JEP remains a work in progress. Significant challenges remain to USCENTCOM's aspirations of conducting seamless effects-based operations across the IE and the spatial domains, but improvement has been noted across all areas while the staff continues to implement and refine the JEP.

U.S. Central Command Has Made Substantial Progress on Integrating Information Through the Adoption of the Joint Effects Process

The processes adopted between 2020 and 2023 engage senior leaders, centralize and elevate planning where appropriate at the GCC level, align OAI to strategic objectives, layer OAI and effects to achieve objectives, promote information sharing across the command, bridge planning between CUOPS and FUOPS, integrate special activities, and involve the "right people, in the right numbers."[1]

The positive attributes of the JEP helped USCENTCOM move from a reactive posture in the IE to a proactive one: In 2020, information was largely an add-on to operations. In 2023, it is a central consideration. The adopted processes have brought coherence to the OAI that USCENTCOM conducts, and component commands are now better aligned to the GCC and with one another.

---

[1] Military officers with expertise in planning and operations, in-person interview with the authors, September 14, 2022; retired senior military officer with expertise in OIE, planning, and operations, telephone interview with the authors, October 5, 2022; military officers with expertise in planning, in-person interview with the authors, December 5, 2022; retired senior military officer with expertise in OIE, planning, and operations, telephone interview with the authors, November 7, 2022.

The Joint Effects Process Moved U.S. Central Command from a Reactive to a Proactive Posture in the Information Environment

For the past 20 years, USCENTCOM was a high-priority theater because of ongoing conflicts. As national policy and strategic goals have changed, USCENTCOM recently became less focused on kinetic warfighting and more focused on assuring partners and deterring adversaries. At the same time, the command has seen a reduction in the number and scale of deployments. This necessitates a mindset shift: Whereas information used to be an additional consideration or an enabler of kinetic operations, it is now the most important consideration.[2] In 2020, however, the command was not well postured for effects-based operations. In 2023, it is apparent that the JEP is supporting effects-based operations better than the previous processes and has elevated the role of information in the command. Particularly, it has engaged senior leaders, led to CONOPS that better layer OAI and effects across domains, promoted collaboration across the command, and integrated special activities with the totality of ongoing operations.[3]

Processes Have Brought Coherence to Operations, Activities, and Investments

USCENTCOM routinely conducts numerous OAI across the theater. Under previous processes, the linkages between the operations USCENTCOM conducted and higher-level objectives was often unclear, as was the relationship between operations.[4] Under the processes that are current at the time of writing, OAI are more-greatly aligned to campaign objectives.[5] The processes also help to generate CONOPS that support multiple objects or to coordinate multiple OAI to achieve a single desired effect.[6] One criticism was that, in 2020, USCENTCOM was "asleep at the wheel," and revisions to the battle rhythm have helped the command take purposeful actions.[7]

---

[2] Military officer with expertise in planning and operations, in-person interview with the authors, December 5, 2022; retired senior military officer with expertise in OIE, planning, and operations, telephone interview with the authors, November 7, 2022.

[3] See Chapter 4, particularly Figure 4.1.

[4] SME, interview with the authors, September 15, 2022; senior military officers with expertise in OIE and planning, in-person interview with the authors, December 5, 2022.

[5] Senior military officers with expertise in OIE and planning, in-person interview with the authors, December 5, 2022; USCENTCOM, 2022c.

[6] Military officer and civilian with expertise in PSYOP and planning, in-person interview with the authors, December 6, 2022; retired senior military officer with expertise in OIE, planning, and operations, telephone interview with the authors, November 7, 2022; USCENTCOM, 2022b; USCENTCOM, 2022c.

[7] Military officers with expertise in planning, in-person interview with the authors, December 5, 2022.

Service Component Commands Are Now Better Aligned to the Geographic Combatant Command and with One Another

The JEP centralizes the appropriate level of planning with the GCC staff. Prior to its implementation, the GCC staff sat atop the components that reported what they were doing.[8] The bulk of operational planning still takes place in the subordinate HQ (as it should), but now the USCENTCOM staff is better able to refine and align their CONOPS to bring coherence to the command's overall campaign. It also fosters collaboration up, down, and across the chain of command by forcing conversations in the key B2C2WG.[9]

The Joint Effects Process Is a Work in Progress, and Areas for Improvement Remain to Fully Integrate Information at U.S. Central Command

While the JEP has helped USCENTCOM to move toward an effects-based approach that better incorporates information, much work remains to be done. Operating in the IE is inherently difficult, and no command will consistently do it perfectly. Therefore, the JEP will always be a work in progress. USCENTCOM faces three general challenges that prevent the command from fully reaching its aspirations for effects-based operations in the IE. These are assessment, intelligence, and personnel.

Assessment of Information Effects Remains a Challenge

USCENTCOM has improved—but still struggles with—its ability to assess operational effectiveness and progress toward the campaign plan. This is partly because defining effects and understanding the relationship between cause and effect in the IE is inherently difficult. A significant portion of these challenges, however, stems from higher-level plans and objectives. Defining effects and indicators of success requires SMART objectives. Many SMEs who we interviewed for this study explained that IMOs are often too vague and abstract to support follow-on planning.[10] The staff also has yet to establish a method of feedback for ongoing planning and execution to inform higher-level plans and objectives.[11] The TCP, IMOs, and the TCO naturally (and perhaps, rightfully) change relatively infrequently. Objectives should

[8] Military officers with expertise in planning and operations, in-person interview with the authors, September 14, 2022; retired senior military officer with expertise in OIE, planning, and operations, telephone interview with the authors, October 5, 2022.

[9] Retired senior military officer with expertise in OIE, planning, and operations, telephone interview with the authors, November 7, 2022; military officers with expertise in planning and operations, in-person interview with the authors, September 14, 2022.

[10] Military officer with expertise in planning and operations, in-person interview with the authors, December 5, 2022.

[11] Military officers with expertise in planning and operations, in-person interview with the authors, September 14, 2022; military officer with expertise in assessment, in-person interview with the authors, September 16, 2022.

not remain entirely static as conditions change and lessons are learned. USCENTCOM is still improving in its ability to refine objectives in a way that promotes better planning and assessment. Also, assessment is a behavior that must be learned and continually practiced. While such tools as the C2IE application have helped staff to define effects and log assessment data, they are only as good as the data entered into them. USCENTCOM continues to work to inculcate a culture of continuous assessment.

Intelligence Is Not Fully Integrated into the Effects Process

Like any military organization, USCENTCOM has always made routine use of intelligence. The JEP has improved USCENTCOM's ability to conduct intelligence-driven operations by including touch points with the J2 in the planning process and by employing human terrain analysts to understand the environment and likely effects.[12] More work remains to truly conduct the types of intelligence-driven operations that USCENTCOM aspires to. By *intelligence-driven*, we mean that intelligence is fully integrated in planning OAI, from defining the intended effect to assessing the impact. As USCENTCOM reorients after two decades of in-theater conflict, it must also reorient its use of intelligence from a primary focus on kinetic strikes to a focus on understanding the behaviors of target audiences and helping planners define effects.[13] Collection requirements could be better defined to support the assessment of OAI because many of these are informational in nature, meaning that expert support to understand themes and changes in behavior is at a premium.

Personnel Are Central to Continued Evolution

A primary reason for the JEP's success is the efforts of USCENTCOM personnel, including the work of dedicated actions officers and the buy-in of key leaders. Personnel turnover still creates challenges in getting the right skill sets in the right places and maintaining collective competence. The ability to think in terms of effects is not inculcated evenly across the U.S. military staff on-hand. Most people who report to the USCENTCOM staff will not be used to focusing on informational effects. The types of analysis the USCENTCOM staff hopes to do—planning effects based on desired behaviors as well as their assessment—require unique blends of talent (both niche skills of specialists and the eclectic background of generalists). For this reason, some SMEs hoped that the command could access personnel with niche skills, such as army strategists, regional experts, and psychological operators. Regarding the military personnel system, turnover has hampered the JEP's evolution and continues to be

---

[12] USCENTCOM, 2022c; senior military officers with expertise in OIE and planning, in-person interview with the authors, December 5, 2022; retired senior military officer with expertise in OIE, planning, and operations, telephone interview with the authors, November 7, 2022; civilians with expertise in social media activities and planning, in-person interview with the authors, December 6, 2022; civilians with expertise in intelligence, in-person interview with authors, September 16, 2022.

[13] Civilian with expertise in assessments, in-person interview with authors, December 6, 2022; senior military officers with expertise in OIE and planning, in-person interview with the authors, December 5, 2022; civilians with expertise in intelligence, in-person interview with authors, September 16, 2022.

a barrier. As the organization learns the effects-based processes, it suffers a setback when members depart and are replaced by new people that have to start learning the processes all over again. Government civilians and contractors have helped in this regard, but this puts the command in the position where they are relying on key personalities and not necessarily building their bench.[14]

Recommendations

Sustain the Joint Effects Process While Continuing to Refine and Improve Processes

Our assessment of the JEP indicates that it has improved USCENTCOM's processes in nearly every relevant area.[15] In some cases, the command performs at an exceptional level. In others, the staff has progressed from having little or no abilities to carrying out these processes at a basic level or with some challenges. Because of this, USCENTCOM operates on an effects-based process and integrates information into its battle rhythm.

USCENTCOM should sustain the JEP and continue refining it. It is clear from our engagements that the command's process is not perfect, but given the complexities of the IE and military operations in general, perfect will always be an aspiration. USCENTCOM should continually assess itself to understand how to sustain its successes and improve in those areas where challenges remain.

Continue to Challenge Habitual Activities; Continue to Connect Operations, Activities, and Investments to Theater Campaign Order and Intermediate Military Objectives

One of the biggest problems that the JEP has helped to address—and one of the persistent stumbling blocks for its continued improvement—is the tendency of the GCC to operate on inertia. Multiple SMEs discussed historical problems with "doing things because we've always done them," "doing things because they seemed like good ideas," or "basing our activities on resources, instead of the other way around."[16] While our interviewees unanimously agreed

---

[14] Retired senior military officer with expertise in OIE, planning, and operations, telephone interview with the authors, November 7, 2022; military officer with expertise in planning and operations, in-person interview with the authors, December 5, 2022.

[15] See the analysis presented in Chapter 4, particularly Figure 4.1.

[16] Retired senior military officer with expertise in OIE, planning, and operations, telephone interview with the authors, November 7, 2022; senior military officers with expertise in OIE and planning, in-person interview with the authors, December 5, 2022.

that these tendencies are reduced, they are all still present to a significant extent.[17] The JEP allows USCENTCOM to start planning from outcomes and to challenge conventional wisdom.[18] The staff should make sure that the JEP is fully leveraged in this way. Senior leaders can inculcate a questioning attitude and frank dialogue by example, ask hard questions, and encourage subordinates to do the same. When the staff sees feedback influencing the process, it creates a virtuous cycle of continual improvement.

A closely related area in which USCENTCOM has made progress but still struggles is connecting OAI to clear effects grounded in the TCO and the IMOs that the command is trying to achieve. As mentioned previously, some of this is challenging because of how higher-level guidance is written.[19] Still, planners now have structures that allow them to craft CONOPS based on effects and to examine ongoing OAI in light of campaign objectives. Keeping the process grounded in the TCP, TCO, and IMOs will help USCENTCOM further develop a campaigning mindset in which activities are conducted to move the command closer to its military objectives through their cumulative effects.

Develop and Implement a Method to Articulate Risk of Action and Risk of Inaction for All Concept Sketches

USCENTCOM personnel described the difficulty of conceptualizing and measuring risk in the IE. In the opinion of some SMEs, the risk of action in the IE is often exaggerated, while the risk of inaction is rarely discussed.[20] There is no right answer when it comes to describing and measuring risk, but USCENTCOM can adapt its processes to include risk in the conversation and move toward a state in which risk is more fully incorporated into planning, especially in the IE.[21]

One way to do this is to provide a basic format to articulate the risk of action and inaction for all CONOPS. For example, products briefed in the IEWG could include a section on risk for the associated OAI. Risk can be considered on the front end but also carried through the process. When upcoming and ongoing OAI are briefed in the JEWG, planners can continue to discuss the associated risks and how to mitigate them. When a priority effect reaches the

[17] Senior military officers with expertise in OIE and planning, in-person interview with the authors, December 5, 2022; retired senior military officer with expertise in OIE, planning, and operations, telephone interview with the authors, November 7, 2022; civilian with expertise in assessments, in-person interview with authors, December 6, 2022.

[18] Senior military officers with expertise in OIE and planning, in-person interview with the authors, December 5, 2022.

[19] Military officer with expertise in planning and operations, in-person interview with the authors, December 5, 2022.

[20] Retired senior military officer with expertise in OIE, planning, and operations, telephone interview with the authors, November 7, 2022.

[21] Military officers with expertise in planning, in-person interview with the authors, December 5, 2022.

JEB, senior leaders will be better prepared to make decisions understanding the risks of both action and inaction, which have been refined by the staff along with the rest of the concept.

Adopt an Effects Template to Increase Process Discipline

One way to nurture effects-based thinking—while increasing process discipline and improving assessment—could be to adopt an *effects template*. By providing a general (and flexible) structure for component, JTF, and USCENTCOM staff to use, USCENTCOM can encourage the type of planning desired in its battle rhythm events while capturing the data needed to enable both decisionmaking and assessment. This template is not meant to be an administrative item or an end in itself. While it can certainly provide a simple, digestible format for slides and information papers, it is more about the mental process the staff should go through when developing a CONOPS.[22]

All concept sketches and assessment should include the following:

- desired effects (provided by the organization proposing the OAI)
- the specific "focus IMO" that the effect is grounded in (with input from J39, J35, and J5)
- the estimated time for the anticipated change in behavior or the environment
- a narrative assessment (before and after, with input from J39 and public affairs officer).

This template serves as a guideline for how action officers should plan and how senior leaders should digest the information presented to them. It establishes the basic content and flow for the briefings and conversations in each of the battle rhythm events and a simple outline for staff products. This flexible framework allows the staff to prioritize substance over form, and outcome over process, while giving staff a basic structure to shape their thinking and interaction.

Improve Assessment

The most significant challenges revealed by RAND's case studies and discussions with SMEs were related to assessment. This will always be a difficult area for military HQ, and especially for a GCC's given the scale and breadth of their operations. However, there are actions that USCENTCOM can take in the near-term to improve how the staff and subordinate components assess operations. These fall into three general recommendations: improving IMOs,

[22] The effects template is not meant as a deliverable or due out but as an enabling constraint to amplify effects-based thinking. For short introductions to enabling constraints, see Trent Hone, "Leadership Teams and Enabling Constraints," webpage, February 9, 2016; and Cynefin.io, "Constraints," wiki page, March 17, 2022. For scholarly articles related to this concept in different contexts, see Darja Šmite, Nils Brede Moe, Marcin Floryan, Javier Gonzalez-Huerta, Michael Dorner, and Aivars Sablis, "Decentralized Decision-Making and Scaled Autonomy at Spotify," *Journal of Systems & Software*, No. 200, February 15, 2023; and Slawomir Czech, "Institutions as Enabling Constraints. A Note on Social Norms, Social Change and Economic Development," *Economics and Law*, Vol 13, No. 2, June 2014.

defining supporting intelligence collection requirements, and designing assessment during the planning process.

Continue to Improve Intermediate Military Objectives

First and foremost, USCENTCOM should reexamine the TCP and TCO, particularly how they develop and communicate IMOs. Multiple SMEs pointed to the vague and abstract nature of IMOs as the starting point for difficulties with assessment.[23] This is only logical: If operations are to be grounded in effects, poorly articulated objectives will lead to poorly articulated effects, which will drive OAI that may or may not be aligned to the intent of the plan and cohere with one another.[24] The effects of these OAI will be difficult to understand because they were not clearly defined in the first place. In contrast, clear IMOs make it easier to explain what specific changes in behavior USCENTCOM is trying to achieve. The more precise planners are with the behavioral changes they desire, the more relevant the indicators of success are, and the more feasible it is to rank them from most to least favorable. After USCENTCOM conducts an OAI, component, JTF, and GCC staff can look for the indicators of success that they selected and assess whether those are present and, if so, to what degree. This is not to say that assessment will ever be straightforward or easy, but any amount of improvement in the clarity of objectives will lead to tangible downstream improvements.

Define Supporting Intelligence Collection Requirements

Cognitive effects are about changing behavior. To assess changes in behavior, one must observe whether and to what extent their predefined indicators are present. This depends on a baseline of what behaviors were present before the OAI (whether it is historical military deployment patterns, public opinion in different areas or among different communities, or the disposition of key leaders). It also depends on sensing and sensemaking within the operating environment to identify the indicators and determine how much has changed from the baseline. This all depends on intelligence. For the types of assessments USCENTCOM wishes to conduct, it will need additional intelligence support to plan and assess effects.[25]

The JEP has made some progress in this area (for example, by bringing intelligence briefers into the JEWG and leveraging human terrain professionals), but much work remains to be

---

[23] Military officer with expertise in assessment, in-person interview with the authors, September 16, 2022; military officer with expertise in planning and operations, in-person interview with the authors, December 5, 2022.

[24] For more on the importance of objectives as the foundation for assessment and how to work to improve assessment by improving objectives, see Christopher Paul, Jessica Yeats, Colin P. Clarke, and Miriam Matthews, *Assessing and Evaluating Department of Defense Efforts to Inform, Influence, and Persuade: Desk Reference*, RAND Corporation, RR-809/1-OSD, 2015.

[25] For more on the challenges related to intelligence support for OIE and overcoming them, see Schwille et al., 2020.

done.[26] The J39 needs access to the necessary resources and processes by which it can shape collection requirements over time to establish relevant baselines and the specific requirements that link to indicators of success. Assessment data will improve if indicators of success are defined based on a strong baseline and the command has the tools to collect relevant data on those indicators. This will most often be a qualitative—and sometimes subjective—endeavor. However, the more intelligence is integrated into the process, the less leaders have to rely on their own assumptions about the relationship between what they did and what they are able to observe.

Include Assessment During Planning for Short, Medium, and Long Time Horizons

Assessment should be "baked into" the planning process.[27] As discussed, when planners define an effect to be achieved, they should also include how they will know whether this effect has occurred and begin to develop concrete indicators of success. This looks different according to the time horizon. Because the JEP helps bridge the gap between near- and long-term planning, assessment needs to be tailored to the short, medium, and long term. For example, the J39 is improving its ability to assess effects and should continue to expand and add rigor to this process. The J5 thinks in terms of campaign assessment and should continually assess progress toward the IMOs and campaign plan. This is only possible if assessment is included when defining effects in the first place and is conducted at the appropriate level during execution. For example, a military deployment might be linked to one or two IMOs, with multiple effects defining each. Each of these effects might have a distinct set of indicators that would be observable in different time frames. These could be things like near- and medium-term changes in the military posture of regional actors or what world leaders are saying in the press. In the longer term, this deployment would be part of the U.S. military's overall presence in the region, and the relevant indicators would change if the overall deployment patterns are, for example, intended to assure regional partners.

Sustain and Expand Use of the Command and Control in the Information Environment Software

C2IE helps USCENTCOM track OAI in time and space and has improved the data available for assessments. As discussed, the system is only as useful as the data entered into it. If USCENTCOM continues to refine the JEP, push back against habitual activities, develop

[26] USCENTCOM, 2022c; senior military officers with expertise in OIE and planning, in-person interview with the authors, December 5, 2022; retired senior military officer with expertise in OIE, planning, and operations, telephone interview with the authors, November 7, 2022; civilians with expertise in social media activities and planning, in-person interview with the authors, December 6, 2022; civilians with expertise in intelligence, in-person interview with authors, September 16, 2022.

[27] For practical advice on planning and conducting assessment of OIE, see Paul et al., 2015.

an effects template, and improve assessment, C2IE will only grow in value. USCENTCOM should continue to use the system as it refines the JEP because it helps provide structure to the changes the command is trying to make. It is important to remember that IT systems are not solutions in and of themselves but a means of sorting and communicating information. As long as the JEP's continued evolution remains focused on substance, C2IE and related tools can be helpful to inculcating an effects-based process.

Integrate Key Leader Engagement as a Capability into the Joint Effects Process

KLE is important OAI. The USCENTCOM staff is making progress on linking the commander's engagements to clear effects and to other OAI.[28] Challenges remain, because KLE planning is often process- and protocol-oriented (making sure the visit happens and goes well), and key themes are often handled by the front office and Commander's Action Group (CAG). There is a KLEWG, but planning touch points are not present, and the process is not fully integrated into the rest of the JEP.[29] Ideally, J5 plans and J3 operations (the J39 in particular) would have a role in shaping KLE. Those in charge of logistics and protocol (the front office and CAG) would be involved in battle rhythm events (particularly the JEWG) where they could help socialize their objectives and shape other OAI. There is a natural tendency to plan such things as talking points together with logistics and protocol from a process standpoint, but broader possible effects should be considered and incorporated. Additionally, KLE readouts need to be distributed at the appropriate level of dissemination to key players to better shape the JEP. Ideally, KLE as OAI are planned in an effects-based way and cohere with the other OAI undertaken across USCENTCOM.

Suggestions for Further Research

RAND's study of USCENTCOM's JEP reveals several potential avenues for future research. Chief among these are tailoring intelligence to understand behavior, improving campaign assessment, developing a framework for the analysis and presentation of risk, and assessing the operational effectiveness in the IE.

Behavioral Intelligence for Effects-Based Operations

Effects-based planning requires an understanding of the incentive structures for target audiences and their underlying attitudes. It also depends on an established baseline for the target audience's behavior to measure effects against. Intelligence is also important to assessment

[28] Military officer with expertise in OIE and planning, in-person interview with the authors, September 15, 2022.

[29] Military officers with expertise in planning, in-person interview with the authors, December 5, 2022.

because indicators of success must be observed to understand the effect they have on the environment. There are undoubtedly intelligence professionals across the U.S. military that understand this dynamic, including at USCENTCOM. It is unclear, however, whether personnel at USCENTCOM and across the joint force are sufficiently trained and educated to think in these terms. It is also unclear whether USCENTCOM's current organization and processes optimally incorporate intelligence for effects-based planning and assessment.

Here, we offer the term *behavioral intelligence* as a starting point. While this encompasses the invaluable work done by human terrain analysts, it can also refer to the analysis of political-military decisionmaking or military activity. We do not mean to suggest that a new intelligence discipline is needed, but existing intelligence disciplines could be better applied to an effects-based paradigm to help plan operations and assess them. Further research is needed regarding how analysts should be trained and developed, both in their career progression and at their given command. There is also room for deeper inquiry into the role of intelligence in USCENTCOM's battle rhythm and in processes across GCCs and joint HQ more broadly.

Campaign Assessment at U.S. Central Command and in the Joint Force

Campaign assessment is one of the most challenging areas for USCENTCOM. Further inquiry could help the command improve its ability to measure progress against the campaign plan. Additionally, as DoD seeks to inculcate a campaigning mindset across DoD and the joint force, there is room for further analysis to understand what this looks like in practice.

At the command level, campaign assessment could be studied through a workshop that includes staff and outside SMEs. One approach might be selecting actual IMOs and effects as exemplars and working through them. Lessons learned from this process could be captured in a product that the command could then apply to other IMO, effects, and assessment challenges. Across the joint force, there is opportunity to apply other qualitative methods to understand how different commands at the operational and strategic level conduct campaign assessment. Literature reviews, focus groups, workshops, and interviews could generate a body of work that identifies common challenges, emergent practices, and other insights that organizations like USCENTCOM and the broader DoD could use to improve campaign assessment.

Develop a Framework for the Analysis and Presentation of Risk

Earlier in this chapter, we recommended that USCENTOM develop and implement an approach to articulating the risk of action and inaction related to all concept sketches. We recognize, however, that fully developed frameworks for the discussion of risks related to information and OIE do not yet exist. We understand that there are pockets of promising practice in certain places in the joint force, among select allies, and in certain sectors in industry. Thoughtful research would plumb existing efforts in this area, synthesize with existing risk assessment frameworks for other LOEs, and match to the distinctive context of the USCENTCOM AOR.

Improving Operational Assessment for Operations in the Information Environment

USCENTCOM SMEs spoke about the difficulty in assessing the effectiveness of activities in the IE. Our research showed a tendency to define adequate MOP, but significant problems defining clear effects and useful indicators of success and establishing cause and effect. As USCENTCOM and the broader joint force become less concerned with bombs and more concerned with tweets (to paraphrase an interviewee), they will need new ways to measure success in the IE.

Existing RAND research could be leveraged to improve assessment practices at USCENTCOM.[30] Furthermore, tools related to understanding and analyzing narratives could be examined across the government and private sectors, along with their potential cross-context applicability to different kinds of military operations. Literature reviews, interviews, and field observation could examine how different military organizations are thinking about effects in the IE and produce insights that USCENTCOM or another military entity could use to inform how it assesses the impacts of their operations. Lastly, wargaming tools—such as command-level or interagency tabletop exercises, emulations, or simulations—could be applied to generate insights about operational effects in the IE in the context of a specific command and in military operations more broadly.

[30] See Paul et al., 2015; Christopher Paul, *Assessing and Evaluating Department of Defense Efforts to Inform, Influence, and Persuade: Worked Example,* RAND Corporation, RR-809/4-OSD, 2017.

Abbreviations

AOR	area of responsibility
B2C2WG	boards, bureaus, centers, cells, and working groups
C2IE	command and control in the information environment
CAG	Commander's Action Group
CCJ3	combatant command J3 (operations)
CCP	U.S. Central Command campaign plan
CDR	commander
CONOPS	concept of operations, concepts of operations
CSWG	Commander's Synchronization Working Group
CUOPS	current operations
DCDR	deputy commander
DoD	U.S. Department of Defense
FUOPS	future operations
GCC	geographic combatant command
GFM	global force management
GFMAP	global force management and allocation process
HQ	headquarters
IE	information environment
IEWG	influence effects working group
IMO	intermediate military objective
IAWG	inter-agency working group
IOWG	information operations working group—defunct
ISR	intelligence, surveillance, and reconnaissance
JCMB	joint collection management board
JEB	joint effects board
JEP	joint effects process
JEWG	joint effects working group
JFE	joint fires element
JMWC	Joint Military Information Support Operations Web Operations Center
JTF	joint task force
JTWG	joint targeting working group
KLE	key leader engagement
KLEWG	key leader engagement working group
LOE	line of effort

MEWG	MISO Effects Working Group
MILDEC	military deception
MISO	military information support operations
MOE	measures of effectiveness
MOP	measures of performance
OAI	operations, activities, and investments
OIE	operations in the information environment
OPSEC	operations security
OPT	operational planning team
PPB&E	policy, programing, budgeting, and execution
PRAC	planning and resources alignment conference
PSYOP	psychological operations
SAOB	Special Activities Oversight Board
SAOC	Special Activities Oversight Council
SAWG	special activities working group
SCC	service component command
SMART	specific, measurable, achievable, relevant, and time bound
SME	subject-matter expert
STO	special technical operations
TCO	theater campaign order
TCP	theater campaign plan
TDWG	target development working group
TF	task force
USCENTCOM	U.S. Central Command
USSOCOM	U.S. Special Operations Command
WebOps	web operations
WG	working group

References

Adkins, Keith, and Tom Evans, "Information as a Joint Function: A Doctrinal Perspective," briefing slides, Joint Information Operations Proponent, Joint Staff J39 Strategic Effects Division, Deputy Director for Global Operations, May 4, 2018. As of July 10, 2023: https://www.jcs.mil/Portals/36/Documents/Doctrine/jdpc/12_Info_Joint_Function04May18v2_1.pptx?ver=2018-05-29-122024-197

Alberts, David S., John J. Garstka, Richard E. Hayes, and David A. Signori, *Understanding Information Age Warfare*, CCRP, 2001.

Boyd, John, "An Organic Design for Command and Control," in Grant T. Hammond, ed., *A Discourse on Winning and Losing*, Maxwell Air Force Base, March 1, 2018.

CENTCOM J39, "Joint Effects Working Group Charter," briefing, MacDill Air Force Base, September 7, 2021.

Cynefin.io, "Constraints," wiki page, March 17, 2022.

Czech, Slawomir, "Institutions as Enabling Constraints. A Note on Social Norms, Social Change and Economic Development," *Economics and Law*, Vol. 13, No. 2, June 2014.

Davis, Paul K., *Effects-Based Operations (EBO): A Grand Challenge for the Analytical Community*, RAND Corporation, MR-1477-USJFCOM/AF, 2001. As of July 2023: https://www.rand.org/pubs/monograph_reports/MR1477.html

Demus, Alyssa, Elizabeth Bodine-Baron, Caitlin McCulloch, Ryan Bauer, Christopher Paul, Jonathan Fujiwara, Benjamin J. Sacks, Michael Schwille, Marcella Morris, and Kelly Beavan, *Operationalizing Air Force Information Warfare*, RAND Corporation, RR-A1740-1, forthcoming.

Glavy, Matthew G., and Eric X. Schaner, "21st-Century Combined Arms: Gaining Advantage Through the Combined Effects of Fires, Maneuver, and Information," *Marine Corps Gazette*, September 2022.

Gormley, Bill, "James Q. Wilson, *Bureaucracy: What Government Agencies Do and Why They Do It*," in Martin Lodge, Edward C. Page, and Steven J. Balla, eds., *The Oxford Handbook of Classics in Public Policy and Administration*, Oxford, 2015.

Hinote, Clint, *Centralized Control and Decentralized Execution: A Catchphrase in Crisis?* Air Force Research Institute, March 2009.

Hone, Trent, "Leadership Teams and Enabling Constraints," webpage, February 9, 2016. As of July 2023: https://trenthone.com/2016/02/09/leadership-teams-and-enabling-constraints/

Joint Doctrine Note 1-19, *Competition Continuum*, U.S. Joint Chiefs of Staff, June 3, 2019.

Joint Publication 1-0, *Joint Personnel Support*, U.S. Joint Chiefs of Staff, December 1, 2020.

Joint Publication 3-04, *Information in Joint Operations*, U.S. Joint Chiefs of Staff, September 14, 2022.

Joint Publication 3-0, *Joint Operations*, U.S. Joint Chiefs of Staff, incorporating change 1, October 22, 2018.

Joint Publication 3-33, *Joint Task Force Headquarters*, U.S. Joint Chiefs of Staff, January 31, 2018.

Joint Publication 3-61, *Public Affairs*, U.S. Joint Chiefs of Staff, incorporating change 1, August 19, 2016.

Joint Publication 5-0, *Joint Planning*, U.S. Joint Chiefs of Staff, December 1, 2020.

Joint Staff J7, *Insights and Best Practices Focus Paper: Design and Planning*, 1st ed., July 2013.

Joint Staff J7, *Insights and Best Practices Focus Paper: Joint Headquarters Organization, Staff Integration, and Battle Rhythm*, 3rd ed., September 2019.

Libicki, Martin C., Brian A. Jackson, David R. Frelinger, Beth E. Lachman, Cesse Cameron Ip, and Nidhi Kalra, *What Should Be Classified? A Framework with Application to the Global Force Management Data Initiative*, RAND Corporation, MG-989-JS, 2010. As of July 2023: https://www.rand.org/pubs/monographs/MG989.html

Marine Corps Doctrinal Publication 8, *Information*, U.S. Department of the Navy, June 21, 2022.

Paul, Christopher, *Assessing and Evaluating Department of Defense Efforts to Inform, Influence, and Persuade: Worked Example*, RAND Corporation, RR-809/4-OSD, 2017. As of July 2023: https://www.rand.org/pubs/research_reports/RR809z4.html

Paul, Christopher, "Is It Time to Abandon the Term Information Operations?" *Strategy Bridge*, March 11, 2019.

Paul, Christopher, and Richard C. Baffa, *Intelligence Support for Operations in the Information Environment: Dividing Roles and Responsibilities Between Intelligence and Information Professionals*, RAND Corporation, RR-3161-EUCOM, 2020. As of July 2023: https://www.rand.org/pubs/research_reports/RR3161.html

Paul, Christopher, Colin P. Clarke, Bonnie L. Triezenberg, David Manheim, and Bradley Wilson, *Improving C2 and Situational Awareness for Operations in and Through the Information Environment*, RAND Corporation, RR-2489-OSD, 2018. As of July 2023: https://www.rand.org/pubs/research_reports/RR2489.html

Paul, Christopher, and Isaac Porche III, "Toward a U.S. Army Cyber Security Culture," *International Journal of Cyber Warfare and Terrorism*, Vol. 1, No. 3, July–September 2012.

Paul, Christopher, Michael Schwille, Michael Vasseur, Elizabeth M. Bartels, and Ryan Bauer, *The Role of Information in U.S. Concepts for Strategic Competition*, RAND Corporation, RR-A1256-1, 2022. As of July 2023: https://www.rand.org/pubs/research_reports/RRA1256-1.html

Paul, Christopher, Jessica Yeats, Colin P. Clarke, and Miriam Matthews, *Assessing and Evaluating Department of Defense Efforts to Inform, Influence, and Persuade: Desk Reference*, RAND Corporation, RR-809/1-OSD, 2015. As of July 2023: https://www.rand.org/pubs/research_reports/RR809z1.html

Paul, Christopher, Jessica Yeats, Colin P. Clarke, Miriam Matthews, and Lauren Skrabala, *Assessing and Evaluating Department of Defense Efforts to Inform, Influence, and Persuade: Handbook for Practitioners*, RAND Corporation, RR-809/2-OSD, 2014. As of July 2023: https://www.rand.org/pubs/research_reports/RR809z2.html

Schrage, Michael, "Like It or Not, You Are Always Leading by Example," *Harvard Business Review*, October 5, 2016.

Schwille, Michael, Anthony Atler, Jonathan Welch, Christopher Paul, and Richard C. Baffa, *Intelligence Support for Operations in the Information Environment: Dividing Roles and Responsibilities Between Intelligence and Information Professionals*, RAND Corporation, RR-3161-EUCOM, 2020. As of July 2023: https://www.rand.org/pubs/research_reports/RR3161.html

Schwille, Michael, Jonathan Welch, Scott Fisher, Thomas M. Whittaker, and Christopher Paul, *Handbook for Tactical Operations in the Information Environment*, RAND Corporation, TL-A732-1, 2021. As of July 2023:
https://www.rand.org/pubs/tools/TLA732-1.html

Šmite, Darja, Nils Brede Moe, Marcin Floryan, Javier Gonzalez-Huerta, Michael Dorner, and Aivars Sablis, "Decentralized Decision-Making and Scaled Autonomy at Spotify," *Journal of Systems & Software*, No. 200, February 15, 2023.

Thomson, Scott K., and Christopher E. Paul, "Paradigm Change: Operational Art and the Information Joint Function," *Joint Force Quarterly*, No. 89, 2nd Quarter 2018.

USCENTCOM—*See* U.S. Central Command.

U.S. Central Command, *CENTCOM Joint Effects Board Information Flow*, March 15, 2022a, Not available to the general public.

U.S. Central Command, *CENTCOM Joint Effects Working Group*, September 1, 2022b, Not available to the general public.

U.S. Central Command, *CENTCOM Joint Effects Working Group*, September 15, 2022c, Not available to the general public.

U.S. Joint Chiefs of Staff, *Joint Publication Joint Concept for Operating in the Information Environment (JCOIE)*, July 25, 2018.